50 Quantum Physics Ideas
You Really
Need to Know

Joanne Baker

greenfinch

Contents

Introduction

The story of quantum physics has as many twists and turns as it has strange phenomena. A stream of vivid characters – from Albert Einstein to Richard Feynman – have puzzled over the interiors of atoms and the nature of forces over the past century. But physics has trumped even their wild imaginations.

The quantum world runs according to the physics of the very small. But subatomic goings-on are hardly clockwork, and are often baffling. Elementary particles pop in and out of existence and once-familiar substances like light seem impossible to pin down, behaving like waves on one day and a stream of bullets the next.

The more we have learned, the stranger the quantum universe has become. Information can be 'entangled' between particles, raising the possibility that everything is connected by invisible threads. Quantum messages are transmitted and received instantaneously, breaking a taboo that no signal can exceed light speed.

Quantum physics is not intuitive – the subatomic world behaves quite differently from the classical world that we are familiar with. The best way to understand it is to follow the path of its development, and to grapple with the same puzzles that the pioneers of the theory wrestled with.

The first chapters summarize how the field emerged at the dawn of the twentieth century, when physicists were starting to dissect the atom and comprehend the nature of light. Max Planck introduced the term 'quanta', arguing that energy came in small packets rather than a continuum. The idea was applied to the structure of the atom, where electrons orbited a compact nucleus in shells.

Out of that work grew quantum mechanics, with all its paradoxes. As particle physics gathered pace, quantum field theories and the standard model emerged to explain it. Finally the book explores some of the implications – for quantum cosmology and concepts of reality – and highlights recent technological developments, such as quantum 'dots' and quantum computing.

Joanne Baker

01 Energy conservation

Energy powers movement and change. It is a shape shifter that takes many forms, from heat given off in burning wood to the speed gained by water flowing downhill. It may swap from one type to another. But energy is never created or destroyed. It is always conserved overall.

The idea of energy as the cause of transformations was familiar to the ancient Greeks – *energeia* means 'activity' in Greek. We know that its magnitude scales with the force we apply and the distance by which an object subjected to it shifts. But energy is still a slippery concept for scientists. It was in investigating the nature of energy that the ideas of quantum physics originated.

When we push a supermarket trolley, it rolls along because we are giving it energy. The trolley is being powered by the chemicals combusted in our bodies, transmitted by the force of our muscles. When we throw a ball we also convert chemical energy into motion. The Sun's heat comes from nuclear fusion, where atomic nuclei are crushed together, and give out energy in the process. Energy appears in many guises: from speeding bullets to lightning strikes. But its origin can always be traced back to another kind. Gunpowder created the bang of the gun. Molecular motions stirred up the static electricity in a cloud that was released in the vast spark. When energy changes from one type to another it makes matter move or change. Because it simply changes form, energy is never created or destroyed. It is conserved: the total amount of energy in the universe, or any completely isolated system, stays the same.

Conservation

In ancient Greece, Aristotle was the first to realize that energy seemed to be conserved, although he had no means of proving it. It took centuries for early scientists (then known as natural philosophers) to understand the different forms of energy individually, and then to link them together.

Galileo Galilei experimented in the early 17th century with a swinging pendulum. He noticed that there was a balance between how fast the bob moved in the centre of its swing and how high it climbed at the end. The higher the bob was released, the faster it swung in between, rising to around the same height at the end. Over the full cycle, energy was being exchanged from 'gravitational potential' (associated with height above the ground) to 'kinetic' (speed) energy.

The 17th-century mathematician Gottfried Leibniz referred to energy as 'vis viva', or life force. The physicist polymath Thomas Young introduced the word energy in the sense we use now in the early 19th century. But exactly what energy is has remained elusive.

Although it acts on vast bodies, from a star to even the whole universe, in its essence energy is a small-scale phenomenon. Chemical energy arises as atoms and molecules rearrange their structures during reactions. Light and other forms of electromagnetic energy are transmitted as waves, which interact with atoms. Heat reflects molecular vibrations. A compressed steel spring withholds elastic energy within its structure.

Energy is intimately tied to the nature of matter itself. Albert Einstein, in 1905, revealed that mass and energy are equivalent. His famous $E = mc^2$ equation states that the energy (E) released by the destruction of a mass (m) is m times the speed of light (c) squared. Because light travels at 300 million metres per second (in empty space), crushing even a few atoms releases an enormous quantity of energy. Our Sun and nuclear power stations release energy in this way.

Other rules

Properties linked to energy can also be conserved. Momentum is one. Linear momentum, the product of mass and velocity, is a measure of how hard it is to slow down a moving body. A heavy supermarket trolley has more momentum than an empty one, and is difficult to stop. Momentum has a direction as well as a size, and both aspects are conserved together. This is put to good effect in snooker – if you hit a stationary ball with a moving one, the final paths of both will sum to give the velocity and direction of the first moving ball.

Momentum is also conserved for rotating objects. For an object spinning about a point, angular momentum is defined as the product of the object's linear momentum and its distance from the point. Ice skaters conserve angular momentum when they spin. They whirl slowly when their arms and legs are outstretched; they speed up by pulling their limbs in to their body.

Another rule is that heat always spreads from hot to cold bodies. This is the second law of thermodynamics. Heat is a measure of atomic vibration, so atoms jiggle more and are more disordered within hot bodies than in cooler ones. Physicists call the amount of disorder, or randomness, 'entropy'. The second law states that entropy always increases, for any closed system with no external influences.

How do refrigerators work then? The answer is that they create heat as a by-product – as you can feel if you put your hand near the back. Fridges don't bust the second law of thermodynamics but work with it, creating more entropy by warming the air than they extract for cooling. On average, taking both the fridge and air molecules into account, entropy increases.

> It is just a strange fact that we can calculate some number and when we finish watching nature go through her tricks and calculate the number again, it is the same.
>
> Richard Feynman,
> *The Feynman Lectures on Physics* (1961)

Many inventors and physicists have tried to think of ways of confounding the second law, but none has succeeded. Schemes for perpetual-motion machines have been dreamt up, from a cup that drains and refills itself to a wheel that propels its own rotation by dropping weights along spokes. But when you look closely at their workings they all leak energy – to heat or noise, say.

The Scottish physicist James Clerk Maxwell in the 1860s devised a thought experiment that could create heat without a rise in entropy – although it has never been made to work without an external power source. Maxwell imagined two adjoining boxes of gas, both at the same temperature, linked by a small hole. If one side is warmed, the particles in that side move faster. Normally, a few of them would pass through the hole into the other side, gradually evening out the temperature.

But Maxwell imagined that the reverse could be possible – by some mechanism, which he pictured as a tiny demon or devil that sorted the molecules (known as 'Maxwell's demon'). If such a mechanism could be devised, it could shift fast molecules from the colder side into the hotter box, violating the second law of thermodynamics. No means of doing this has ever been discovered, so the second law prevails.

Ideas and rules about how to move and share energy around, coupled with increasing knowledge of atomic structure, led to the birth of quantum physics in the early 20th century.

The condensed idea
Shape-shifting energy

02 Planck's law

By solving the problem of why coals glow red and not blue, the German physicist Max Planck started a revolution that led to the birth of quantum physics. Seeking to describe both light and heat in his equations, he apportioned energy into small packets, or quanta, and in the process explained why so little ultraviolet light is given off by hot bodies.

It's winter and you're cold. You imagine the cosy glow of a roaring fire – the red coals and the yellow flames. But why do coals glow red? Why does the tip of an iron poker placed within the fire also become red-hot? Burning coals reach hundreds of degrees Celsius. Volcanic lava is hotter, approaching 1,000°C. Molten lava glows more fiercely and can appear orange or yellow, as does molten steel at the same temperature. Tungsten light-bulb filaments are even hotter. With temperatures of thousands of degrees Celsius, similar to the surface of a star, they shine white.

Black-body radiation

Bodies give off light at progressively higher frequencies as they are heated. Especially for dark materials such as coal and iron – which are efficient at absorbing and giving off heat – the spread of frequencies radiated at a particular temperature has a similar form, known as 'black-body' radiation. Most light energy radiates around one 'peak' frequency, which scales with temperature from red towards blue. Energy also leaks out to either side, rising in strength towards the peak at low frequencies, and declining above it. The result is an asymmetric hill-shaped spectrum, known as a 'black-body curve'.

Colour temperature

The colour of a star gives away its temperature. The Sun, at 6,000 kelvins, appears yellow, while the cooler surface of the red giant Betelgeuse (in the constellation Orion) has a temperature of half that. The scorching surface of Sirius, the brightest star in the sky, which shines blue-white, reaches 30,000 kelvins.

A glowing coal might put out most of its light in the orange range, but it also gives off a little low-frequency red and some higher-frequency yellow, but barely any blue. Hotter molten steel shifts this pattern up in frequency, to emit mostly yellow light, with some orange-red and a touch of green.

The ultraviolet catastrophe

By the late 19th century, physicists knew of black-body radiation and had measured its frequency pattern. But they could not explain it. Different theories could describe part of the behaviour but not all of it. Wilhelm Wien concocted an equation that predicted the rapid dimming at blue frequencies. Meanwhile, Lord Rayleigh and James Jeans explained the rising red spectrum. But neither formula could describe both ends.

Rayleigh and Jeans's rising spectrum solution was particularly problematic. Without a means of curtailing its growth, their theory predicted an infinite release of energy at ultraviolet and shorter wavelengths. This problem was known as the 'ultraviolet catastrophe'. The solution came from the German physicist Max Planck, who was trying to unify the physics of heat and light at the time. Planck liked to think mathematically and to tackle physics problems from scratch, starting from the basics. Fascinated by the fundamental laws of physics, notably the second law of thermodynamics and Maxwell's equations of electromagnetism, he set about proving how they were linked.

Quanta

Planck faithfully manipulated his equations, without worrying about what those steps might mean in real life. To make the mathematics easier to work with, he devised a clever trick. Part of the problem was that electromagnetism is described in terms of waves. Temperature on the other hand is a statistical phenomenon, with heat energy shared out among many atoms or molecules. So Planck decided to treat electromagnetism in the same way as thermodynamics. In place of atoms, he envisaged electromagnetic fields as being carried by tiny oscillators. Each one could take a certain amount of the electromagnetic energy, which was shared out among many of these elementary entities.

Planck scaled the energy of each oscillator with frequency, such that $E = h\nu$, where E is energy, ν is light frequency, and h is a constant factor now known as Planck's constant. These units of energy were called 'quanta', from the Latin for 'how much'.

In Planck's equations, quanta of high-frequency radiation have correspondingly high energies. Because the total amount of energy available is capped, there couldn't be many high-energy quanta in the system. It's a bit like economics. If you have \$99 in your wallet, it's likely that there are more bills of smaller denominations than large ones. You could have nine dollar bills, four or more ten-dollar bills but only one 50-dollar bill, if you're lucky. Similarly, high-energy quanta are rare.

Planck worked out the most likely energy range for a set of electromagnetic quanta. On average, most of the energy was midway – explaining the peaked shape of the black-body spectrum. Planck published his law in 1901. It was received with great acclaim as it neatly solved the troublesome 'ultraviolet catastrophe' problem.

Max Planck (1858–1947)

At school in Munich, Germany, Max Planck's first love was music. When he asked a musician where he should go to study it he was told he'd better do something else if he had to ask that question. He turned to physics, but his professor complained that physics was a complete science: nothing more could be learned. Fortunately, Planck ignored him and went on to develop the concept of quanta. Planck endured the deaths of his wife and two sons killed in the world wars. Remaining in Germany, he was able to rebuild physics research there in the aftermath. Today, Germany's prestigious Max Planck research institutes are named after him.

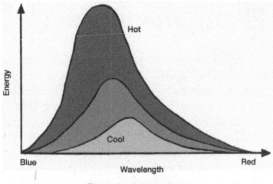

Black-body curves

Planck's concept of quanta was entirely theoretical – the oscillators weren't necessarily real but were a useful mathematical construction to match the physics of waves and heat. But coming at the beginning of the 20th century, a time when our understanding of light and the atomic world was advancing rapidly, Planck's idea had implications beyond anything he imagined. It became the root of quantum theory.

Planck's legacy in space

The most accurately known black-body spectrum comes from space. A faint microwave glow with a precise temperature of 2.73 K emanates from all directions in the sky. Its origin is in the very early universe, a hundred thousand years after the Big Bang when the first hydrogen atoms formed. Heat energy from that time has since cooled as the universe has expanded, and now peaks in the microwave part of the spectrum, following a black-body law. This cosmic microwave background radiation was detected in the 1960s but mapped in detail in the 1990s by NASA's COBE (COsmic Background Explorer) satellite. Europe's latest microwave background mission, launched in 2009, is named after Planck.

The condensed idea
Energy economics

03 Electromagnetism

L ight is an electromagnetic wave. Extending beyond the familiar spectrum of visible light, electromagnetic disturbances range from radio to gamma rays. Now understood as one phenomenon that unites electricity and magnetism, electromagnetism is one of four fundamental forces. Its essence has been the stimulus for both relativity and quantum physics.

We take light for granted, but there is a lot that we don't understand about it. We see shadows and reflections – it doesn't pass through or bounce off opaque or shiny materials. And we know it breaks up into the familiar rainbow spectrum when it passes through glass or raindrops. But what is light really?

Many scientists have tried to answer that question. Isaac Newton showed in the 17th century that each hue of the rainbow – red, orange, yellow, green, blue, indigo, violet – is a fundamental 'note' of light. He mixed them together to produce intermediate shades, such as cyan, and recombined them all into white light, but he could not dissect the spectrum further with the equipment he had. Experimenting with his lenses and prisms, Newton found that light behaves like water waves – bending around obstacles and reinforcing or cancelling where waves overlap. He reasoned that light was made up, like water, of tiny particles, or 'corpuscles'.

We now know that this is not strictly so. Light is an electromagnetic wave, made of oscillating electric and magnetic fields coupled together. But there is more to the tale. In the early 1900s Albert Einstein showed that there are situations where light does behave like a stream of particles, now called photons, which carry energy but have no mass. The nature of light remains a conundrum, and has been central to developments in relativity and quantum theory.

The spectrum

Each of light's hues has a different wavelength, or spacing between adjacent wave crests. Blue light has a shorter wavelength than red; green lies in between. The frequency is the number of wave cycles (peaks or troughs) per second. When a beam of white light passes through a prism, the glass bends (refracts) each colour by a different

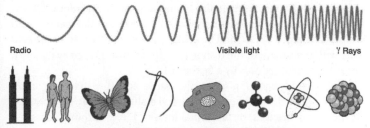

Radio	Visible light	γ Rays

Electromagnetic waves range in wavelength from
thousands of metres to billionths of a metre.

angle, so that red bends least and blue most. As a result the colours
spread out into a rainbow. But the colours don't end there.

Visible light is just one part of the electromagnetic spectrum,
which stretches from radio waves with wavelengths spanning
kilometres to gamma rays with wavelengths much smaller than an
atom. Visible light's wavelength is around a billionth of a metre, close
to the sizes of many molecules. Beyond red light wavelengths of
millionths of a metre is infrared light. At millimetre to centimetre
wavelengths we find microwaves. Short of violet light lie ultraviolet,
X- and gamma (γ) rays.

Maxwell's equations

Electromagnetic waves combine electricity and magnetism. In the
early 19th century, experimenters such as Michael Faraday saw that
these fields could switch from one sort to the other. Moving a magnet
near a wire pushes charges around and causes electricity to flow in
that wire. A changing current passed through a wire coil produces a
magnetic field that can induce a current in another coil – this is the
basis for the electrical transformer, used to scale currents and voltages
for domestic energy.

The big breakthrough came when the Scottish physicist James
Clerk Maxwell managed to encapsulate all this behaviour in just four
equations – known as Maxwell's equations. Maxwell explained how
electricity and magnetism arise from one phenomenon:
electromagnetic waves, comprising an electric field varying like a sine
wave in one direction, accompanied by a magnetic field varying
similarly but oriented at right angles.

Maxwell's first equation is also known as Gauss's law, after the 19th-century physicist Carl Friedrich Gauss. It describes the electric field around a charged object, and how, like gravity, the strength of the field falls with the square of the distance. So if you move twice as far away, the electric field reduces by a factor of 4.

The second equation does the same for the magnetic field. Magnetic (and electric) fields are often visualized by drawing contours of their field strength, or tangential lines of force. Around a magnet, the second law states that these magnetic field lines are always closed loops, travelling from the north to the south pole. In other words, all magnetic field lines must start and end somewhere, and all magnets must have a north and a south pole – there is no such thing as a magnetic 'monopole'. Chopping a bar magnet in half will always recreate north or south poles. Both poles are retained no matter how many times you slice the magnet.

> In order to understand the nature of things, men must begin by asking, not whether a thing is good or bad, noxious or beneficial, but of what kind it is?
> James Clerk Maxwell, 1870

The third and fourth of Maxwell's equations describe electromagnetic induction, the creation and interchange of electric and magnetic forces by moving magnets and currents flowing through wire coils. The third equation describes how varying currents cause magnetic fields, and the fourth how varying magnetic fields create electric currents. Maxwell also showed that light waves, and all electromagnetic waves, travel at a constant speed in a vacuum of around 300 million metres per second.

Encapsulating so many phenomena in a few elegant equations was an enormous feat. Einstein rated Maxwell's achievement on a par with Newton's grand description of gravitation, and applied Maxwell's ideas within his relativity theories. Einstein went a step further and explained how magnetism and electricity were manifestations of the same electromagnetic force seen in different situations. Someone viewing an electric field in one frame would see it as a magnetic field in another frame that was moving relative to the first. But Einstein didn't stop there. He also showed that light isn't always a wave – it can sometimes act as a particle.

James Clerk Maxwell (1831–79)

Born in Edinburgh, Scotland, James Clerk Maxwell became fascinated with the natural world through spending time in the Scottish countryside. At school he was given the nickname 'dafty' because he was so engrossed in his studies. His reputation at first Edinburgh and later Cambridge University was as a bright if disorganized student.

After graduating, Maxwell took Michael Faraday's earlier work on electricity and magnetism and combined it into four equations. In 1862, he showed that electromagnetic waves and light travel at the same speed, and eleven years later he published his four equations of electromagnetism.

The condensed idea
Colours of the rainbow

04 Young's fringes

When a beam of light is split into two, the different trains can mix to either reinforce or cancel the signal. Like water waves, where crests meet, the waves combine and bright bands appear; where peaks and troughs cancel, there is darkness. This behaviour, called interference, proves that light acts like a wave.

In 1801 the physicist Thomas Young shone a beam of sunlight through two very fine slits cut in a piece of card. The light spread out into its constituent colours. But it did not form one classic rainbow, nor even two. To his surprise, on the screen appeared a whole series of rainbow stripes. These are known today as Young's fringes.

What was going on? Young closed off one of the slits. A single broad rainbow appeared, much as you'd expect from shining white light through a prism. The main rainbow was flanked by a few fainter flecks on either side. When he reopened the second slit, the pattern broke up once more into the array of vivid bands.

Young realized that light was behaving just like water waves. Using glass tanks filled with water, he had studied how waves pass around obstacles and through gaps. When a parallel series of waves goes through an opening, such as a harbour entrance in a sea wall, some of them go straight through. But the waves that graze the wall's edge are deflected – or diffracted – into arcs, spreading wave energy to either side of the gap. This behaviour could explain the single-slit pattern. But what about the double-slit fringes?

Throwing a pebble into a lake generates rings of expanding ripples. Throw another stone close to the first, and the two sets of ripples overlap. Where two crests or two troughs meet, the waves combine and grow larger. When a crest and a trough meet, they cancel each other out. The result is a complex pattern of peaks and lows arranged around 'spokes' of flat water.

This effect is known as interference. What happens when the wave grows in size we call 'constructive interference'; its diminution is 'destructive interference'. The size of the wave at any point depends on the difference in the 'phase' of the two interfering waves, or the relative distance between the peaks of each. The same behaviour appears in all waves, including light.

By using a double slit, Young had made two trains of light – one from each – interfere. Their relative phases were dictated by their different paths through and beyond the card. Where the waves combined to reinforce one another, a bright stripe resulted. Where they cancelled, the background was dark.

Huygens' principle

In the 17th century, the Dutch physicist Christiaan Huygens had devised a practical rule of thumb – known as Huygens' principle – for predicting the progression of waves. Imagine freezing a circular ripple for a moment. Each point on that ring can become a new source of circular waves. Each new ripple again becomes a set of new sources. By carrying out this sequence again and again the wave's evolution can be followed.

All that you need to track the wave is a pencil and paper and a pair of compasses. Start by drawing the first wavefront, then use the compasses to describe further circles along it. The wave's next envelope can be worked out by drawing a smooth line through the circles' outer edges. The method is simply repeated.

This easy technique can be applied to follow the trajectories of waves through gaps and around objects placed in their path. In the early 19th century, the French physicist Augustin-Jean Fresnel extended Huygens' principle to more complex circumstances, such as waves encountering obstacles and crossing the paths of other waves.

When waves travel through small gaps, their energy spreads out to either side – through a process called diffraction. Using Huygens' approach, the wave energy sources at the edge of the gap radiate circular ripples, making the wave look almost semicircular after it has gone through. Similarly, waves may diffract energy around corners.

Young's experiment

When Young shone white light through one slit, most of the waves passed through, but diffraction at the edges of the slit produced two close sets of circular waves, which interfered, giving the faint extra fringes at the side of the main bright line.

The amount of diffraction depends on the width of the slit relative to the wavelength of the light going through. The spacing of the side fringes scales with wavelength and inversely to slit width. So a

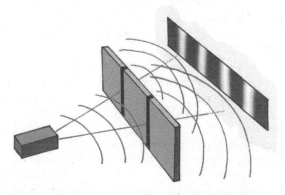

Light waves combine or cancel as they pass through two slits.

narrower slit produces more widely spaced extra fringes, and red light is more spread out than blue.

When a second slit is introduced the result is a combination of the above pattern and a second diffraction pattern from the interference of waves from each of the slits. Because the difference between those sources is much larger than the width of one slit, the fringes that result are more narrowly spaced.

This is what Young saw – many fine fringes due to interference of wave trains through both slits superimposed on a broad fringe pattern due to diffraction through one slit.

Young's discovery was important at the time, because it was contrary to Newton's earlier idea that light was made up of particles, or 'corpuscles'. Because light beams could interfere, Young clearly showed that light is a wave. Particles would have passed straight through the card and built up two stripes on the screen.

But it is not so simple. Physicists since have showed that light is fickle: in some circumstances it behaves as a particle, in others as a wave. Variants of Young's double-slit experiment – passing very faint beams of light or opening and closing slits quickly while the light travels through – are still very important for investigating the nature of light. Some of the weirder findings have fed into tests of quantum theory.

> Each time a man stands up for an ideal . . . he sends forth a tiny ripple of hope, and crossing each other from a million different centres of energy and daring, those ripples build a current that can sweep down the mightiest walls of oppression and resistance.
> Robert Kennedy, 1966

The condensed idea
Wave mixing

05 Speed of light

L ight travels at the same speed whether emitted by a lamp on a bicycle, a fast train or a supersonic jet. Albert Einstein showed, in 1905, that nothing can travel faster than the speed of light. Time and space distort when approaching this universal speed limit. Near light speed, objects become heavier and shorten and time slows.

When you watch an electric storm, the rumble of thunder follows the flash of lightning. The further away the storm is, the greater the thunder's delay. This is because sound travels much more slowly through air than light does. Sound is a pressure pulse; it takes several seconds to cover a kilometre. Light is an electromagnetic phenomenon, and vastly quicker. But through what sort of medium does it move?

In the late 19th century, physicists supposed that space was filled with some sort of electric gas, or 'ether', through which light travels. In 1887, however, a famous experiment proved that this medium did not exist. Albert Michelson and Edward Morley devised an ingenious means to detect the possible movement of Earth as it orbited the Sun against the fixed background of the ether.

In their lab, they fired two beams of light at right angles to one another, reflecting them back off identical mirrors displaced by exactly the same distance. Where the beams met, interference fringes were produced. If Earth moved along the direction of one of the arms, the planet's velocity should add to or subtract from the light's speed against the ether. There would be a difference in how long it took the light to traverse each of the arms, just as a swimmer takes different times to swim across a river with or against a current. As a result, the fringes should move slightly to and fro during a year. But they did not move. The light beams always returned to their starting points at exactly the same time. No matter how or where Earth shifted in space, the speed of light was unchanged. The ether did not exist.

Light always travels at the same speed: 300 million metres per second. This is odd compared with water and sound waves, which may slow down in different media. Plus, in our experience, velocities normally add or subtract – an overtaking car seems to crawl ahead. If you were to shine a torch at the other driver, the beam would travel at the same rate no matter how fast each of you was going. The same is true in a fast train or jet plane.

Einstein and relativity

Why is the speed of light fixed? This question led Albert Einstein to devise his theory of special relativity in 1905. Then employed as a patent clerk in Bern, Switzerland, Einstein worked on physics in his spare time. He tried to imagine what two people, travelling at different speeds, would see if they shone a torch at one another. If the speed of light can't change, Einstein reasoned, then something else must change to compensate.

What changes is space and time. Following ideas developed by Hendrik Lorentz, George Fitzgerald and Henri Poincaré, Einstein made the fabric of space and time stretch so that fast-moving observers would still perceive light's speed as constant. He treated the three dimensions of space and one of time as aspects of a four-dimensional 'space-time'. Speed is

> space is not a lot of points close together; it is a lot of distances interlocked.
> Sir Arthur Stanley Eddington, 1923

distance divided by time, so since nothing can exceed the speed of light, distance must shrink and time must slow to compensate. A rocket travelling away from you at near light speed looks shorter and experiences time more slowly than you do.

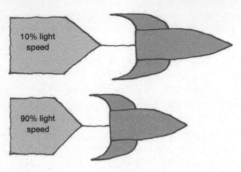

Lengths shorten when travelling near light speed.

Einstein's theory stated that all motion is relative: there is no privileged viewpoint. If you are sitting on a train and see another train next to you moving, you may not know which one is still and which is pulling out of the station. Similarly, although Earth is moving around the Sun and across our own galaxy, we don't perceive that motion. Relative motion is all we can experience.

The flying clocks

Near the speed of light, Einstein predicted that time would slow down. Moving clocks may run at different speeds. This surprising fact was proven in 1971. Four identical atomic clocks were flown twice around the world, two to the east and two to the west. When they arrived back, their times were compared with another identical clock that had stayed on the ground. The moving clocks each lost a fraction of a second compared with the stationary clock, confirming Einstein's special relativity theory.

The velocity of light is to the Theory of Relativity as the elementary quantum of action is to the Quantum Theory: it is its absolute core.
Max Planck, 1948

Objects also grow more massive as they approach light speed, according to $E = mc^2$ (energy = mass × speed of light squared). This weight gain is tiny at slow speeds, but becomes infinite at light speed, such that any further acceleration is impossible. So nothing can exceed the speed of light. And anything with a mass can never quite get there but only approach it, becoming heavier and more difficult to accelerate the closer to light speed it gets. Light itself is made of photons that have no mass, so are unaffected.

Einstein's special relativity caused consternation and took decades to be accepted. The implications, including the equivalence of mass and energy, time dilation and mass, were profoundly different from anything considered before. Perhaps the only reason that relativity was entertained at all was that Max Planck heard about it and became fascinated. Planck's championing of special relativity theory catapulted Einstein into mainstream academia and eventually public fame.

The condensed idea
Everything is relative

06 Photoelectric effect

A series of experiments in the 19th century showed that the wave theory of light was insufficient. Light shone onto a metal surface proved to dislodge electrons, whose energies could only be explained if light was made of distinct 'photon' bullets and not waves.

In 1887, the German physicist Heinrich Hertz was toying with sparks whilst trying to build an early radio receiver. Electricity sent crackling between two metal balls in the transmitter could trigger another spark in a second pair in the receiver – making up a device called a spark-gap generator.

The second spark was ignited more easily, he found, when the receiver spheres were close together – usually just a millimetre or so apart. But, strangely, sparks also caught more readily when the apparatus was bathed in ultraviolet light. That made little sense. Light is an electromagnetic wave, whose energy could have been passed to the electrons in the surface layer of the metal, setting them free in the form of electricity. But further investigations showed that was not so.

Hertz's assistant Philipp Lenard broke down the spark-gap generator into its basic form: two metal surfaces placed in an evacuated glass tube. The plates inside were separated but connected outside the jar by a wire and ammeter to form an electric circuit. Lenard shone light of different brightness and frequency onto the first plate, whilst keeping the second dark. Any electrons knocked out of the first plate would fly across and hit the second, completing the circuit and causing a tiny current to flow. Lenard found that bright light produced more electrons than faint light, as expected, given that more energy was being shone on the plate. But varying the intensity of the light had barely any effect on the speed of the electrons knocked out. Both bright and faint sources produced electrons with the same energy, which he measured by applying a slight opposing voltage to stop them. This was unexpected – with the greater energy input by intense light he expected to find faster electrons.

Colours of light

Other physicists turned to the problem, including the American Robert Millikan. Testing beams of different colours, he found that red

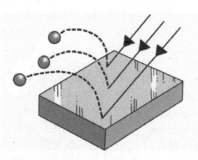

Blue light kicks electrons out of metals

light could not dislodge any electrons, no matter how bright the source. Yet ultraviolet and blue light worked fine. Different metals had a different 'cutoff frequency', below which light shone on them could not release electrons. The energy (speed) of the electrons emitted above this threshold scaled with the frequency of the light. The gradient of that relationship is known as Planck's constant.

This behaviour was startling: according to the ideas of the day, light waves should work in the opposite way. Electromagnetic waves flooding the metal surface should slowly boil off electrons. Just as storm waves impart more energy than small ripples, the stronger the light, the more energetic and numerous the electrons dislodged should be. Nor should frequency have much effect – in terms of energy imparted to a sitting electron there should be little difference between many small ocean waves and a few large ones. Instead, tiny rapid ripples were readily kicking out electrons, whereas slow swell, no matter how monstrous the waves, left them unmoved.

Another puzzle was that the electrons were being dislodged too quickly. Rather than taking a time to steadily absorb the light energy, even at low light levels

Fifty years of conscious brooding have brought me no closer to answer the question, 'What are light quanta?' Of course today every rascal thinks he knows the answer, but he is deluding himself.
Albert Einstein, 1905

electrons jumped out instantaneously. By analogy, one tiny ripple sent the electron flying. All in all, something must be wrong with the simple electromagnetic wave picture of light.

It seems to me that the observation associated with black-body radiation, fluorescence, the photoelectric effect, and other related phenomena associated with the emission or transformation of light are more readily understood if one assumes that the energy of light is discontinuously distributed in space.

Albert Einstein, 1905

Einstein's photon bullets

In 1905, Albert Einstein explained the weird properties of the photoelectric effect with a radical idea. He won the Nobel Prize for this work in 1921. Drawing on Max Planck's concept of energy quanta, Einstein argued that light exists in tiny energy packets. The quanta of light were later named 'photons'.

Einstein suggested that it was the force of individual photon bullets that knocked electrons out of the metal. Although having no mass, each photon carries a certain amount of energy, in proportion to its frequency. Blue and ultraviolet photons therefore pack more of a punch than red ones. This could explain why the bumped electron's energy also scales with the light's frequency and not with its brightness. A red photon will not dislodge any electron as it doesn't carry enough energy to do so. But a blue photon's clout will kick an electron out. With even more energy, an ultraviolet photon would knock out a faster electron. Adjusting the brightness won't help. Just as firing a grape won't deflect a cannon ball, so increasing the number of impotent red photons will not shift electrons. The immediacy of the effect can be explained – travelling at light speed, all it takes is one photon to kick out the electron.

Millikan's oil-drop experiment

In 1909 Robert Millikan and Harvey Fletcher used a droplet of oil to measure the electric charge of an electron. By suspending it between two charged metal plates, the pair showed that the force needed to keep it aloft always involved a multiple of a basic quantity of electric charge, which they measured to be 1.6×10^{-19} coulombs. This, they supposed, is the charge on one electron.

Einstein's idea of light quanta initially went down like a lead balloon. Physicists didn't like it because they revered the wave description of light summarized so neatly in Maxwell's equations. But a flurry of experiments confirming that the released electrons' energies scaled with the frequency of light quickly turned the crazy idea into a fact.

The condensed idea
Photon bullets

07 Wave–particle duality

At the dawn of the 20th century, the idea that light and electricity were transmitted as waves and that solid matter was made of particles broke down. Experiments revealed that electrons and photons underwent diffraction and interference – just like waves. Waves and particles are two sides of the same coin.

Einstein's 1905 proposal that light energy was transmitted in packets of energy – photons – and not continuous waves was so controversial it took nearly two decades and many further tests to be accepted. At first it seemed to reopen a polarized debate from the 17th century about what light was made of. In fact it heralded a new understanding of the relation between matter and energy.

In the 1600s Isaac Newton had argued that light must consist of particles, as it travelled in straight lines, was cleanly reflected and slowed in 'refractive' materials such as glass. Christiaan Huygens and, later, Augustin-Jean Fresnel showed that light must be a wave, because of the way it bent around obstacles, diffracting, reflecting and interfering. James Clerk Maxwell cemented the wave theory in the 1860s in his four equations summarizing electromagnetism.

> There are two sides to every question.
> Protagoras, 485–421 BC

Einstein's proposition that light was made of particles rocked the boat. But beyond that it set up an uncomfortable tension that is still with us today. For light is not either a wave or else a particle – it is both. And the same is true of other electromagnetic phenomena.

The pursuit of light

Light behaves as a stream of torpedoes under some circumstances, such as in the photoelectric effect apparatus, and as waves in others, as in Young's double-slit experiment. Whatever we set out to measure it to be, light adjusts its behaviour so that that side of its character comes through in the experiment we subject it to.

Physicists have devised canny experiments to catch light out, and reveal its 'true' nature. None of them has captured its pure essence. Variants of Young's double-slit experiment have pushed the wave–particle duality of light to its limits, but the synergy remains.

Light whose intensity is so dim that individual photons can be
watched going through the slits gives the same interference pattern if
you wait long enough – individual photons pile up to give collectively
the familiar fine fringes. If you close one slit, the fired photons' locus
reverts to a broad diffraction figure. Open the slit again and the stripes
reappear at once.

It is as if the photon is in two places at once, and 'knows' what the
status of the second slit is. It doesn't matter how quick you are, it is
impossible to trick the photon. If one of the slits is shut whilst the
photon is in flight, even after the particle has crossed through the gap
but before it hits the screen, it will behave in the correct way.

The photon behaves as if it goes through both slits simultaneously.
If you try to pin it down, say by placing a detector in one, then
strangely the interference pattern disappears. The photon becomes a
particle when you treat it as one. In every case physicists have tested,
interference fringes will appear or disappear according to how you
treat the photons.

Matter waves

Wave–particle duality doesn't just apply to light. In 1924, Louis-Victor
de Broglie suggested that particles of matter – or any object – could
also behave as waves. He assigned a characteristic wavelength to all
bodies, tiny and large. The larger the object, the smaller the

wavelength. A tennis ball flying across a court has a wavelength of 10^{-34} metres, much smaller than a proton's width. Because macroscopic objects have minuscule wavelengths, too small to see, we cannot spot them behaving like waves.

Three years later de Broglie's idea was confirmed: electrons were found to diffract and interfere like light. Electricity had been known to be carried by particles – electrons – since the late 19th century. Just as light didn't need a medium in which to travel, in 1897 J.J. Thomson showed that electrical charge could traverse a vacuum, so a particle must be necessary to carry it. This did not sit comfortably with the belief that electromagnetic fields were waves.

In 1927 at Bell Labs in New Jersey, Clinton Davisson and Lester Germer fired electrons at a crystal of nickel. The electrons that emerged were scattered by the atomic layers of the crystal lattice and the outgoing beams mingled to produce a recognizable diffraction pattern. Electrons were interfering, just as light did. The electrons were behaving like waves. A similar technique was also being used to ascertain the structures of crystals by firing X-rays through them – X-ray crystallography. Although their origin was uncertain when discovered by Wilhelm Conrad Röntgen in 1895, X-rays were soon realized to be a high-energy form of electromagnetic radiation.

In 1912, Max von Laue realized that the short wavelengths of X-rays were comparable with atomic spacings in crystals, so if fired through the layers they would diffract. The crystal's geometry could be calculated from the positions of the bright spots that result. This method famously proved the double-helix structure of DNA in the 1950s.

A related experiment proved Einstein's photon concept in 1922. Arthur Compton succeeded in scattering X-rays off electrons, measuring the small change in their frequency that resulted – known as the Compton effect. Both X-ray photons and electrons were behaving like billiard balls. Einstein was right. Moreover all electromagnetic phenomena behaved like particles.

For matter, just as much as for radiation, in particular light, we must introduce at one and the same time the corpuscle concept and the wave concept.
Louis de Broglie, 1929

Today, physicists have witnessed wave–particle behaviour in neutrons, protons and molecules, even large ones such as the microscopic carbon footballs known as 'buckyballs'.

The condensed idea
Two sides of the same coin

08 **Rutherford's atom**

In the late 1800s, physicists began to break into the atom. First they revealed electrons, and then a hard core, or nucleus, of protons and neutrons. To explain what bound the nucleus together, a new fundamental force – the strong nuclear force – was proposed.

Atoms were once thought to be the smallest building blocks of matter, but little more than a century ago, all that changed. Physicists started to dissect the atom and showed it to be made of many layers, like a Russian doll. The first layer was that of the electron. Firing an electric current through gas contained in a glass tube, the Englishman Joseph John (J.J.) Thomson freed electrons from atoms in 1887.

He knew little of how they were distributed in matter, and proposed the simple 'plum-pudding model' of the atom, where negatively charged electrons were sprinkled like prunes or raisins through a dough of positive charge. The attraction between the electrons and positive charges supposedly held the atom together as they were mixed throughout the 'pudding'.

The deeper layers were the targets of an experiment in 1909. Ernest Rutherford performed an intriguing test with his colleagues Hans Geiger and Ernest Marsden. With the aim of testing the plum-pudding model, they fired heavy alpha particles – a form of radiation emanating from radium or uranium – through a very thin gold foil, just a few atoms thick.

They expected that most of the alpha particles would pass straight through. In fact, a small proportion of the particles (one in several thousand) bounced right off the foil. Many reversed direction, being deflected by large angles (90 to 180 degrees), as if they had hit something hard, like a baseball bat. Rutherford realized that within the gold atoms that made up the foil lay some compact, hard and massive cores.

Naming the nucleus

Thomson's soft plum-pudding model could not explain this. It viewed an atom as a jumble of positive and negative charges, none of which was hard or heavy enough to block the alpha particle. Rutherford concluded that the gold atoms must have a dense centre. He called it the 'nucleus', after the Latin word for the kernel of a nut. It was the dawn of nuclear physics, the physics of the atomic nucleus.

Physicists and chemists knew about the masses of different elements through the periodic table. In 1815, William Prout had suggested that atoms were composed of multiples of the simplest atom – hydrogen. But this could not explain the weights of the elements easily. The second element, helium, for example was not double but four times heavier than hydrogen.

Just over a century later, in 1917, Rutherford showed that other elements do contain hydrogen nuclei – the positively charged particles were given off when alpha particles (helium nuclei) were fired through nitrogen gas, which was turned into oxygen in the process. This was the first time one element had been deliberately transmuted into another. To avoid confusion with hydrogen gas itself, in 1920 Rutherford named the bare hydrogen nucleus the 'proton', after the Greek for 'first'.

> It was almost as incredible as if you fired a 15-inch shell at a piece of tissue paper and it came back to hit you.
>
> Ernest Rutherford, 1936

Components of the nucleus

To explain atomic weights, Rutherford imagined that the nucleus was made up of some number of protons, plus a few electrons within it to partially balance the charge. The rest of the electrons sat outside the nucleus in shells. Hydrogen, the lightest element, has a nucleus

Most of an atom's mass resides in its nucleus.

containing just one proton with one electron orbiting it. Helium, he reasoned, would have four protons and two electrons in its nucleus – to give the double positive charge of an alpha particle – with two more orbiting outside.

The concept of nuclear electrons quickly turned out to be false. In 1932 a new particle was found by Rutherford's associate James Chadwick. A neutral particle with the same mass as the proton, it was heavy enough to free protons from paraffin but had no charge. It was named the neutron, and the model of the atom was rearranged.

Atomic weights could be explained by a mix of neutrons and protons in the nucleus. A carbon-12 atom, for instance, contains six protons and six neutrons in the nucleus (to give it a mass of 12 atomic

Carbon dating

A heavy form of carbon is used for dating archaeological artefacts, such as wood or charcoal from fires, over some thousands of years. Carbon's normal weight is 12 atomic units, but it occasionally appears in a form with 14 units. Carbon-14 is unstable, and decays radioactively. The time it takes for half the atoms to decay by emitting a beta particle, to become nitrogen-14, is 5,730 years. This slow reaction can be used for dating.

units) and six orbiting electrons. Alternative forms of elements with odd weights are called isotopes.

The nucleus of an atom is minuscule. Just a few femtometres (10^{-15} metres, or one ten million billionth of a metre) across, the atom's core is a hundred thousand times more compact than the electron orbits that surround it. That ratio is equivalent to the length of Manhattan, around ten kilometres, relative to the Earth's diameter.

The nucleus is also heavy and dense – virtually all the atom's mass, possibly comprising many tens of protons and neutrons, is crammed into that tiny region. But how can all those positively charged protons stick together? Why don't they repel and blow the nucleus apart? Physicists needed a new sort of force to glue the nucleons together, which they called the strong nuclear force.

The strong force acts on scales so small that it is only important within the nucleus. Outside it is much weaker than the electrostatic force. So if you could grab two protons and push them together, at first you would feel their repulsion. Keep pushing, though, and they would snap together like building blocks. If you compressed them further, the protons would not budge. So protons and neutrons are bound tightly within the nucleus, which is compact and hard.

I am a great believer in the simplicity of things and as you probably know I am inclined to hang on to broad and simple ideas like grim death until evidence is too strong for my tenacity.

Ernest Rutherford, 1936

With gravity, electromagnetism and the weak nuclear force, the strong force is one of four fundamental forces.

The condensed idea
The compact core

09 Quantum leaps

Electrons circle the nucleus in shells of different energy, like the orbits of the planets. Niels Bohr described how electrons may hop between the shells, and as they do so emit or absorb light corresponding to the energy difference. These jumps are known as quantum leaps.

In 1913, the Danish physicist Niels Bohr improved upon Rutherford's model of the atom by establishing how electrons were arranged around the nucleus. Bohr imagined that the negatively charged electrons travelled in orbits around a positively charged core, just as planets orbit the Sun. He also explained why their orbits lie at particular distances from the centre, tying in atomic structure with quantum physics.

> Everything we call real is made of things that cannot be regarded as real.
>
> Niels Bohr

Electrons are held in place around the nucleus by electrostatic forces – the mutual attraction of positive and negative charges. But moving charges, he knew, should lose energy. Just as moving electrical current generates a field around a wire or in a radio transmitter, moving electrons give off electromagnetic radiation.

Early theories of the atom therefore predicted that orbiting electrons should shed energy and slowly spiral into the nucleus, giving off electromagnetic waves of ever higher frequency – a continuous screech of rising pitch. This obviously wasn't happening in reality. Atoms don't spontaneously collapse, and no sign of the high frequency shriek was ever spotted.

Spectral lines

Instead, atoms give off light only at very specific wavelengths. Each element produces a characteristic set of 'spectral lines', like a sort of musical scale in light. Bohr supposed that these 'notes' were related to the energies of the electron orbits. Only in those shells was the electron stable and not subject to the loss of electromagnetic energy.

Electrons, Bohr postulated, can move between orbits and step up and down the scale, as if climbing rungs on a ladder. These steps are known as quantum leaps or jumps. The difference in energy between the rungs is gained or lost by the electron absorbing or emitting light at that corresponding frequency. These produce the spectral lines.

So electrons could take on only certain packets of energy – it was quantized, just as Max Planck had described in his explanation of black-body radiation. The energy differences between the orbits could be expressed as an integer multiple of the frequency of light and a fixed unit – called Planck's constant (h).

The angular momentum of each shell scales so that each subsequent orbit has 1, 2, 3, 4 and so on times that of the first. The integer labels for the different energy states of the electrons are known as the primary 'quantum numbers': $n = 1$ corresponds to the lowest orbit, $n = 2$ to the next, and so on.

In this way Bohr was able to describe the set of energies of the hydrogen atom, the simplest atom, with one electron orbiting a single proton. And those energies fitted the spectral lines of hydrogen well, solving a long-standing puzzle.

Bohr extended his model to heavier atoms, which have more protons and neutrons in their nuclei and more orbiting electrons. He supposed that each orbit could hold only a certain number of electrons and they filled up from the lowest energy upwards. When one level was full then electrons accumulated in higher shells.

Because the outer electrons' view of the nucleus is partly blocked by the inner electrons, they don't feel as strong an attractive force from the centre as they would if they sat alone. Nearby electrons also

Types of chemical bonds

Covalent bond: Pairs of electrons are shared by two atoms.

Ionic bond: Electrons from one atom are removed and attached to another atom, resulting in positive and negative ions that attract each other.

Van der Waal's bond: Electrostatic forces attract molecules in a liquid.

Metallic bond: Positive ions are islands in a sea of electrons.

repel one another. So the energy levels of large atoms differ from those of hydrogen. More sophisticated modern models do a better job than Bohr's original of explaining these differences.

Exploring electron shells

Bohr's shell model explains the different sizes of atoms and how they vary across the periodic table. Those with a few loosely bound outer electrons are able to swell more easily than those with full outer shells. So elements like fluorine and chlorine on the right-hand side of the table tend to be more compact than those on the left side, such as lithium and sodium. The model also explains why noble gases are inert – their outer shells are full and so cannot acquire or donate electrons by reacting with other atoms. The first shell takes just two electrons before it is filled. So helium, with two protons in its nucleus binding two electrons, has a full outer shell and doesn't interact easily. The second shell takes eight electrons – it is full for the next noble gas, neon. Things get more complicated for the third shell and beyond, because the electron orbitals adopt non-spherical shapes. The third shell takes eight electrons, but there is another dumb-bell-like configuration that can take ten more – thereby explaining the transition elements, such as iron and copper.

It is wrong to think that the task of physics is to find out how Nature is. Physics concerns what we say about Nature.

Niels Bohr

The shapes of large electron orbitals go beyond Bohr's simple model and are difficult to calculate even today. But they dictate the forms of molecules, as chemical bonds arise from the exchange of electrons. Bohr's model doesn't work well for very large atoms, like iron. Nor can it explain the strengths or detailed structures of spectral lines. Bohr didn't believe in photons at the time he developed his model, which was based on the classical theory of electromagnetism.

Bohr's model was superseded in the late 1920s by quantum-mechanical versions. These accommodate the wave-like properties of an electron and treat its orbit as a sort of probability cloud – a region of space where there is some probability that the electron is. It is not possible to know exactly where the electron is at a given time.

Electrons can jump from one orbit to another, gaining or losing electromagnetic radiation with a frequency (ν) proportional to the energy difference ($\triangle E$) according to the Planck relation, where h is Planck's constant:

$$\triangle E = E_2 - E_1 = h\nu$$

Yet Bohr's insight remains useful across chemistry today, as it explains a myriad patterns, from the periodic table's structure to hydrogen's spectrum.

The condensed idea
Electron energy ladder

10 Fraunhofer lines

Light can be absorbed or emitted when an electron in an atom moves from one energy level to another. Because electron shells lie at fixed energies, the light only takes on certain frequencies and appears as a series of stripes – called Fraunhofer lines – when split by a prism or multi-slit grating.

Since Isaac Newton shone a beam of sunlight through a glass prism in the 17th century we've known that white light is made up of a mix of the colours of the rainbow. But if you look more closely, the spectrum of sunlight contains many black stripes – as if a bar code has been imprinted. Particular wavelengths are chopped out as the light from the Sun's heart passes through the star's gaseous outer layers.

Each 'absorption line' corresponds to a particular chemical element seen in various states and energies. Common ones are hydrogen and helium, which make up the bulk of the Sun, and products of its burning, including carbon, oxygen and nitrogen. By mapping the pattern of the lines you can work out the chemistry of the Sun.

The English astronomer William Hyde Wollaston spotted dark lines in the solar spectrum in 1802, but the first detailed examination was carried out in 1814 by the German lens maker Joseph von Fraunhofer, after whom they are now named. Fraunhofer managed to list more than 500 lines; modern equipment identifies thousands.

In the 1850s the German chemists Gustav Kirchhoff and Robert Bunsen worked out through experiments in the lab that each element produces a unique set of absorption lines – each has its own bar code. Elements may also emit light at those frequencies. Fluorescent neon lights, for instance, give off a series of bright lines that correspond to the energy levels of atoms in the neon gas within the tube.

The precise frequency of each spectral line corresponds to the energy of a quantum jump between two electron energy levels in a particular atom. If the atom is in a gas that is very hot – such as that within the neon light tube – the electrons try to cool down and lose energy. As they drop to a lower energy level they give off a bright emission line at a frequency corresponding to the energy difference.

Cool gases on the other hand absorb energy from a background light source, boosting an electron into a higher level. This results in a dark absorption line – a gap – in the spectrum of the source behind. The study of spectral chemistry, known as spectroscopy, is a powerful technique for revealing the contents of materials.

Gratings

Rather than using glass prisms, which are limited in their power and are bulky, a device with a row of parallel narrow slits cut into it can be inserted into the light beam. This is called a grating: Fraunhofer made the first from aligned wires.

Gratings are much more powerful tools than prisms and can bend light through broader angles. They also take advantage of the wave properties of light. A beam sent through any one slit spreads out its energy due to diffraction. The angle over which it is skewed scales with the wavelength of the light but inversely to the width of the slit. Very fine slits spread out the light more widely, and red light is deflected more than blue light.

Joseph von Fraunhofer (1787–1826)

Born in Bavaria, Germany, Fraunhofer was orphaned at the age of 11 and became an apprentice glassmaker. In 1801, he narrowly missed being killed when the workshop collapsed and buried him. He was rescued by a prince – Maximilian I Joseph of Bavaria – who supported his education and helped him move to a monastery specializing in fine glassmaking. There he learned how to make some of the best optical glass in the world, and eventually became the institute's director. Like many glassmakers of the time, he died young – aged 39 – owing to poisoning from heavy metal vapours used in the trade.

When two or more slits come into play, interference between the wave trains takes place – peaks and troughs of light waves either reinforce or cancel out, creating a pattern of light and dark stripes, known as fringes, on a screen. This pattern is made up of two effects superimposed: the single-slit pattern appears, but within each of its fringes is a finer series of stripes, whose divisions scale inversely to the distance between the slits.

Gratings are like a bigger version of Young's double-slit experiment. Because there are many slits, the bright fringes are sharper. The more slits, the brighter the fringes. Each fringe is a mini-spectrum. Physicists can build bespoke gratings to dissect the spectrum of light at finer and finer resolution by varying the density and size of the slits. Gratings are widely used in astronomy for observing the light from stars and galaxies, to see what they are made of.

Diagnostics

Although white light spreads out to give a smooth red-blue-green spectrum, atoms emit light only at certain frequencies. This bar code of 'spectral lines' corresponds to the energy levels of electrons within them. The wavelengths for common elements, such as hydrogen, helium or oxygen, are well known from laboratory experiments.

All integral laws of spectral lines and of atomic theory spring originally from the quantum theory. It is the mysterious organon on which Nature plays her music of the spectra, and according to the rhythm of which she regulates the structure of the atoms and nuclei.
Arnold Sommerfeld, 1919

Bright emission lines result when an electron is too hot and loses energy, dropping into a lower energy state and releasing the excess as a photon. Absorption lines are also possible if atoms are bathed with light with the right energy to knock an electron to a higher orbit. Then the bar code appears as dark stripes against a broader background.

The exact frequency of the lines depends on the energetic state of the atoms and whether they are ionized or not – in very hot gases the outer electrons may be stripped off. Because of their sensitivity, spectral lines are used to probe many fundamental aspects of the physics of the gas. The lines are broadened in hot gases by the atom's motions, which becomes a measure of temperature. The relative strengths of different lines may tell us more, such as how ionized the gas is.

But on closer inspection it all gets complicated – the appearance of finer structures in spectral lines tells us more about the nature of electrons, and has been instrumental in dissecting the properties of atoms at the quantum scale.

The condensed idea
A barcode made of light

11 Zeeman effect

When spectral lines are examined closely they break up into finer structure. Experiments in the 1920s showed that this was due to an intrinsic property of electrons called quantum spin. Electrons behave like spinning balls of charge, and interactions with magnetic and electric fields subtly alter their energy levels.

When hot hydrogen glows, it emits a series of spectral lines. These arise when electrons perform a quantum leap, jumping from a high energy level down into a lower one, as they try to cool down. Each line of the hydrogen spectrum corresponds to a particular jump, when the energy between the two electron levels is converted into light with that corresponding frequency.

When the electron drops from the second level to the first, it emits light with a wavelength of 121 nanometres (nm: billionths of a metre), which is in the ultraviolet part of the spectrum. An electron jumping from the third level to the first gives off higher-energy light with a shorter wavelength of 103 nm. From the fourth it is 97 nm. Because electron shells get closer together as they go up in energy, the energy gaps between them decrease. So the lines for drops into a given shell tend to bunch up in wavelength towards the blue end of the spectrum.

The set of spectral lines that results from electrons falling to a particular level is called a 'series'. For hydrogen, the simplest atom and

Sunspot magnetism

In 1908 the astronomer George Ellery Hale observed the Zeeman effect in the light from sunspots, dark regions on the surface of the Sun. The effect disappeared for light from brighter regions, implying that the sunspots had intense magnetic fields. By measuring the separations of the split spectral lines Hale was able to deduce the strength of those fields. He went on to show that there are symmetries in the magnetic polarity of sunspots, with ones on either side of the solar equator behaving in an opposite way, for instance.

the most common element in the universe, the primary ones are named after scientists. The series of transitions to the first shell is known as the Lyman series, after Theodore Lyman, who discovered it between 1906 and 1914. The first spectral line (from level 2 to 1) is named Lyman-alpha, the second (level 3 to 1) Lyman-beta and so on.

The set of jumps into the second level is known as the Balmer series, after Johann Balmer, who predicted them in 1885. Many of these lines lie in the visible part of the spectrum. The set of leaps to the third energy level is the Paschen series, after Friedrich Paschen, who observed them in 1908. Those lie in the infrared.

Further examination showed that these spectral lines were not pure but had fine structure. When seen at really high resolution a hydrogen line was revealed to be not one but two lines, close together. The energy levels of the electrons generating the lines were being split up into multiples.

Silver bullets

In a famous experiment in 1922, Otto Stern and Walther Gerlach fired a beam of silver atoms from a hot oven through a magnetic field. The beam split in two, making two marks on a photographic plate. As well as its ability to be detected with photographic emulsion, Stern and Gerlach had chosen silver for their test because it has a single outer electron. The goal of their experiment was to look at the magnetic properties of electrons.

As the silver electrons pass through the field, they behave

An electron is no more (and no less) hypothetical than a star.
Arthur Stanley Eddington, 1932

like tiny bar magnets, and experience a force that is proportional to the gradient in the external magnetic field. Stern and Gerlach expected that these forces would be randomly oriented – producing a single smudge on their detector plate. Instead the beam divided in two, making two spots. This implied that the electron 'magnets' had just two possible orientations. That was very odd.

Electron spin

But why does an electron gain any magnetism? In 1925 Samuel Goudsmit and George Uhlenbeck proposed that the electron acts like a spinning ball of charge – a property called electron spin. By the rules

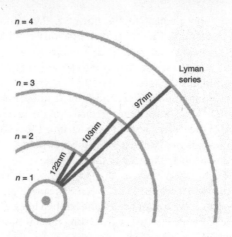

Electrons jumping between energy levels in a hydrogen atom give off light with specific wavelengths. The set of lines resulting from jumps to a particular level is called a series.

of electromagnetism, moving charge generates a magnetic field. The beam in the Stern–Gerlach experiment split in two because the electrons have two directions in which to spin – described as up and down. Those two orientations also explained the fine splitting of the spectral lines – there is a slight energy difference between electrons spinning in the same direction as their orbit, versus those opposing it.

Quantum spin is not really a motion but an intrinsic property of particles. To describe whether the spin is up or down, physicists give electrons and other particles a spin quantum number, which is defined to have a value of plus or minus ½ for electrons.

> There was a time when we wanted to be told what an electron is. The question was never answered. No familiar conceptions can be woven around the electron; it belongs to the waiting list.
> Arthur Stanley Eddington, 1928

Lots of different interactions can arise between spinning electrons and other charged and electromagnetic phenomena – from the electron's own charge and that of the nucleus to external fields. So spectral lines become split in many complex ways.

The splitting of lines arising when electrons lie within magnetic fields is known as the Zeeman effect, after the Dutch physicist Pieter Zeeman. It is seen in the light from sunspots, for example. Line splitting due to an electric field is known as the Stark effect, after Johannes Stark.

The impact of the Stern–Gerlach experiment was huge – it was the first time that the quantum properties of a particle had been laid bare in the lab. Scientists followed it up with more tests, showing for example that the nuclei of some atoms have quantized angular momentum – which also interacts with electron spin to give 'hyperfine' line splitting. And that the electron's spin could be made to switch from one state to another by using varying fields. That finding is now at the root of Magnetic Resonance Imaging (MRI) machines in hospitals.

The condensed idea
Spinning electrons

12 Pauli exclusion principle

No two electrons are the same. Pauli's principle states that each one must have a unique set of quantum properties so that you can tell them apart. This turns out to explain why atoms have certain numbers of electrons in shells, the structure of the periodic table, and why matter is solid even though it's mostly empty space.

In Niels Bohr's 1913 model of the atom, the lowest energy orbital of hydrogen takes just two electrons, the next eight and so on. This geometry is embodied in the block structure of the periodic table. But why are the numbers of electrons per shell limited, and how do electrons know in which energy level to sit?

Wolfgang Pauli sought an explanation. He had been working on the Zeeman effect – the splitting of spectral lines that results when magnetism changes the energy levels of spinning electrons in atoms – and saw similarities in the spectra of alkali metals, having one outer electron, and the noble gases, with full shells. There seemed to be a fixed number of states that electrons could have. This could be explained if every electron had one state, as described by four quantum numbers – energy, angular momentum, intrinsic magnetism and spin. In other words, each electron has a unique address.

Pauli's rule – known as Pauli's exclusion principle – devised in 1925 stated that no two electrons in an atom could have the same four quantum numbers. No two electrons could be in the same place with the same properties at the same time.

Electron organization

Moving along the periodic table to heavier and heavier elements, the number of electrons in an atom grows. The electrons cannot all take the same seat, and they gradually fill higher and higher shells. It is like seats in a tiny cinema filling up, from the screen outwards.

Two electrons may both inhabit the lowest energy state of an atom, but only if their spins are misaligned. In helium, its two electrons can both sit in the lowest shell with opposite spins. In lithium, the third has to bump up to the next shell.

Pauli's rule applies to all electrons and to some other particles whose quantum spin comes in half-integer multiples of the basic unit, including the proton and neutron. Such particles are called 'fermions' after the Italian physicist Enrico Fermi.

Electrons, protons and neutrons are all fermions, so Pauli's exclusion principle applies to the atomic building blocks that make up matter. The fact that no two fermions can sit in the same seat is what gives matter its rigidity. Atoms are mostly empty space, but we can't squeeze them like a sponge or push materials through each other like cheese through a grater. Pauli had answered one of the most profound questions in physics.

The lives of stars

Pauli's principle has implications in astrophysics. Neutron stars and white dwarfs owe their existence to it. When stars bigger than our Sun get old, their nuclear fusion engines falter. They stop converting elements from hydrogen through to iron and become unstable. When the core collapses, the star implodes. Its onion-like layers fall inwards, with some of the gas blown away in a supernova blast.

Wolfgang Pauli (1900–58)

As a precocious schoolboy in Vienna, Wolfgang Pauli smuggled Albert Einstein's papers on special relativity into his desk and secretly read them. Within months of his arrival at university in Munich, Pauli published his first paper on relativity. And then he turned to quantum mechanics. Werner Heisenberg described Pauli as a 'typical nightbird' who spent evenings in cafés and rarely rose to attend morning lectures. Following the suicide of his mother and the failure of his first marriage, Pauli developed a drink problem. Seeking help from the Swiss psychologist Carl Jung, Pauli sent him descriptions of thousands of his dreams, some of which Jung later published. During the Second World War, Pauli moved to the United States for several years, during which he worked hard to keep European science going. He returned to Zurich, receiving the Nobel Prize in 1945.

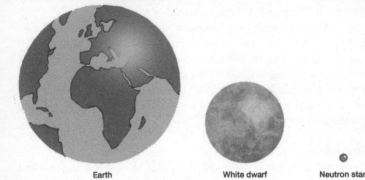

Earth   White dwarf   Neutron star

As the gas collapses, gravity pulls it in further. The embers contract, crushing the atoms together. But the rigid electrons around each atom resist – Fermi's principle holds up the dying star by this 'degeneracy pressure' alone. Such a star is known as a 'white dwarf', and contains about the same mass as the Sun crunched into the Earth's volume. One sugar cube of white dwarf material would weigh a tonne.

For stars much larger than the Sun – above a limit of 1.4 times the solar mass, known as the Chandrasekhar mass limit – the pressure is so great that eventually even the electrons give way. They merge with protons to form neutrons. So the electrons vanish and a 'neutron star' results. Neutrons are also fermions. So they too prop each other up – they cannot all adopt the same quantum state. The remnant star still remains intact but its size drops to a radius of only ten kilometres or so. It is like squashing the Sun's mass into an area the length of Manhattan. A sugar cube's worth of dense neutron star matter would weigh more than a hundred million tonnes. The compaction need not end there – really massive stars eventually become black holes.

Pauli's exclusion principle holds sway right across the universe, from the most basic particle to a distant star.

Bosons

Not every particle is a fermion – some have integer spin. These are called bosons, after the Indian physicist Satyendranath Bose who worked on them. Photons are bosons, as are the particles that carry the other fundamental forces. Some symmetric nuclei can act like bosons, including helium, which is made from two protons and two neutrons. Not subject to Fermi's principle, any number of bosons can have the same quantum properties. So thousands of bosons might act in quantum concert, a phenomenon that is central to odd macroscopic quantum behaviour such as superfluids and superconductivity.

The condensed idea
No two fermions are alike

13 Matrix mechanics

The flood of discoveries about wave–particle duality and the quantum properties of atoms in the 1920s left the field in a quandary. New theories of the atom were needed. The first volley came from the German physicist Werner Heisenberg, who did away with preconceptions about electron orbits and put all the observed variables into one set of matrix-based equations.

In 1920, the Danish physicist Niels Bohr opened a new institute at the University of Copenhagen. Scientists from around the world came to work with him on the atomic theory he was pioneering. Bohr's model of electron orbits explained the spectrum of hydrogen and some properties of the periodic table. But the detailed properties of spectral lines from larger atoms, even helium, didn't fit the theory.

A slew of emerging findings was also challenging Bohr's model of the atom. Evidence for wave–particle duality was multiplying. X-rays and electrons were shown to both diffract and bounce off one another, proving Louis de Broglie's hypothesis that matter could behave like waves and waves like particles. Einstein's idea of the photon nature of light wasn't yet accepted though.

Most physicists, including Bohr and Max Planck, still thought of quantum numbers and rules as arising from regularities in the basic structure of atoms. In the wake of the devastation of the First World War, it was becoming clear that an entirely new sort of understanding of energy quantization was needed.

The German physicist Werner Heisenberg made regular short visits to Copenhagen to study with Bohr, starting in 1924. While working on ways to calculate the spectral lines of hydrogen, Heisenberg came to a realization. Because physicists knew so little about what was really going on within atoms, all there was to work on was what could be observed. He went back to the drawing board and set about devising an intellectual framework that could incorporate all the quantum variables.

Heisenberg suffered badly from hay fever, and in June 1925 he decided to leave his home town of Göttingen and go to stay by the sea, where there was less pollen in the air. He travelled to the small island of Helgoland off the North Sea coast of Germany. It was while staying there that he had his epiphany. It was almost three o'clock in the morning, Heisenberg wrote later, when the final result of his calculations lay before him. At first alarmed by the profound implications of his breakthrough, he became so excited that he could not sleep. He left the house and waited for the sunrise on the top of a rock.

Enter the matrix

What was Heisenberg's revelation? In order to predict the strength of the various spectral lines of an atom, he had replaced Bohr's idea of fixed electron orbits with a mathematical description of them as harmonics of standing waves. He was able to link their properties to quantum jumps in energy, using a set of equations that was equivalent to a series of multiplications. Heisenberg returned to his university department in Göttingen and showed his calculations to a colleague, Max Born. Heisenberg wasn't at all confident, Born recalled later, and referred to his seaside paper as crazy, vague and unpublishable. But Born quickly saw its value.

Born, who had trained extensively in mathematics, saw that Heisenberg's idea could be better written in a shorthand form – as a matrix. Matrices were common in mathematics but had seen little use in physics. A matrix is a table of values on which a mathematical function can be performed on all the entries sequentially. Matrix notation could encapsulate Heisenberg's series of multiplication rules in one equation. With his former student Pascual Jordan, Born condensed Heisenberg's equations into a matrix format. The values in the table linked the energies of the electrons and the lines of the spectrum. Born and Jordan hastily penned a paper and published their work; a third manuscript by the three physicists followed.

Heisenberg's concept was novel because it wasn't obviously based on the picture of electron orbits. And Born and Jordan's concise notation allowed the mathematics to be developed for its own sake. They could now push the theory beyond everyone's preconceptions of what atoms were, and make new predictions.

But 'matrix mechanics' was slow to be picked up and became highly controversial. Not only was it in a weird mathematical language that physicists found unfamiliar, but also there were political hurdles to be overcome among the scientists working in the field. Bohr liked the theory – it related well to his ideas about discrete quantum jumps. But Einstein did not favour it.

Einstein was trying to explain wave–particle duality. Accepting the idea – originally de Broglie's – that electron orbits could be described using the equations of standing waves, Einstein and his followers still hoped that quantum properties could ultimately be described by extending wave theory. But Bohr's followers went in a different direction. The field split in two.

Those scientists who did adopt matrix mechanics pushed it further, using it to explain various quantum phenomena. Wolfgang Pauli, for example, managed to explain the Stark effect – the splitting of spectral lines by an electric field – even though it did not explain his exclusion principle. But the theory could not deal easily with the Zeeman effect and electron spin, and it wasn't compatible with relativity.

> We must be clear that when it comes to atoms, language can be used only as in poetry.
>
> Niels Bohr, 1920 (recollected by Heisenberg)

The uncertainty principle

There were deeper implications that the matrix picture raised too. Because it focused only on energy levels and line intensities, the theory did not, by definition, say anything about where an electron actually was or how it moved at any time. And questions remained about what the numbers in the matrices were and what they meant in real life. Matrix mechanics seemed very abstract.

Because the results of an observation – the energies of electrons and spectral lines – must be real, whatever clever tricks were being used to manipulate the mathematics, everything unreal had to ultimately cancel out. The upshot was that matrix mechanics could not explain some qualities of atoms simultaneously. This led eventually to Heisenberg's 'uncertainty principle'.

> All the qualities of the atom of modern physics are derived, it has no immediate and direct physical properties at all.
> Werner Heisenberg, 1952

But before those issues could be tackled, matrix mechanics was upstaged by a new theory. The Austrian scientist Erwin Schrödinger proposed a competing explanation of electron energies that was based on the equations of waves.

The condensed idea
Quantum times-tables

14 Schrödinger's wave equation

n 1926 Erwin Schrödinger managed to describe the energies of electrons in atoms by treating them not as particles, but as waves. His equation calculates a 'wavefunction' that describes the probability of an electron being somewhere at a given time. It is one of the main foundations of quantum mechanics.

In the early 20th century it was becoming clear that the concepts of particles and waves are closely entwined. Albert Einstein showed in 1905 that light waves could also appear as streams of bullet-like photons, whose energy scaled with the frequency of the light. Louis de Broglie proposed in 1924 that all matter did the same – electrons, atoms and any objects made of them have the potential to diffract and interfere as waves.

In Niels Bohr's 1913 theory of the atom, electrons lived in fixed orbits about the nucleus. Electrons take the form of standing waves – like a guitar string set resonating. In an atom, the electron energies are limited to certain harmonics. A whole number of wavelengths of the electron must fit along the circumference of the electron's orbit.

But how do electrons move? If they are wave-like, then they would spread out over the entire orbit, presumably. If they are compact particles, might they travel along circular paths like planets in orbit around the Sun? How would these orbits be arranged? Planets all occupy one plane. Atoms have three dimensions.

The Austrian physicist Erwin Schrödinger set out to describe the electron mathematically as a three-dimensional wave. Struggling to make progress, in December 1925 he headed off to an isolated cabin in the mountains, with a lover in tow. His marriage was notoriously troubled and he had many girlfriends with his wife's knowledge.

Breakthrough

Schrödinger was an unconventional man – often dishevelled and known for always having his walking boots and a rucksack on him. One colleague recalled how he was occasionally mistaken for a tramp when he was attending conferences.

While at the cabin, Schrödinger's mood lifted. He realized that he had already made a lot of progress on his calculations. He could publish what he had done and then keep working on the more difficult aspects – such as incorporating relativity and time dependence – later.

The 1926 paper that resulted presents an equation describing the likelihood that a particle behaving as a wave is in a certain place, using the physics of waves and probability. Today it is a cornerstone of quantum mechanics.

Mathematics of likelihood

Schrödinger's equation correctly predicted the wavelengths of hydrogen's spectral lines. A month later he submitted a second paper, applying his theory to basic atomic systems, such as the diatomic molecule. In a third paper he pointed out that his wave equation was exactly equivalent to Werner Heisenberg's matrix mechanics and could explain the same phenomena. In a fourth paper he incorporated time dependence, showing how a wavefunction would evolve.

Because Schrödinger's explanation was simple for physicists familiar with classical wave theory to understand, the equation was quickly hailed as revolutionary and immediately overtook

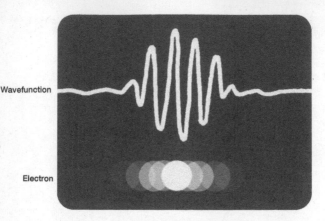

Wavefunction

Electron

Wavefunctions describe the probability of where an electron is. The higher the wavefunction's amplitude the more likely the electron is to be in that place.

Heisenberg's matrix mechanics in the popularity stakes. The matrix theory had fewer fans, as it was expressed in an abstract, unfamiliar sort of mathematics.

Einstein, who favoured the wave approach, was delighted with Schrödinger's break-through. Bohr was interested but still drawn to matrix mechanics, which better described his dislocated quantum leaps. Quantum theory was developing fast, but it was splintered. Were we really learning anything about the real world?

Wavefunctions

Schrödinger expressed the probability of the particle being in a given place at some time in terms of a 'wavefunction', which includes all the information we know about that particle. Wavefunctions are difficult to grasp, as we do not witness them in our own experience and find them hard to visualize and interpret. As with Heisenberg's matrix mechanics, there was still a gulf between the mathematical description of a wave–particle and the real entity, for instance the electron or photon.

God runs electromagnetics by wave theory on Monday, Wednesday and Friday, and the Devil runs them by quantum theory on Tuesday, Thursday and Saturday.
Lawrence Bragg, quoted in 1978

In conventional physics, we would use Newton's laws to describe the motion of a particle. At any instant, we could say exactly where it is and the direction in which it is moving. In quantum mechanics, however, we can only talk about the probability of the particle being in some place at some time.

What might a wavefunction look like? In Schrödinger's equation, a lone particle floating in free space has a wavefunction that looks like a sine wave. The wavefunction is zero in places where the particle's existence can be ruled out, such as beyond the limits of an atom.

The amplitude of the wavefunction can be determined by considering the allowed energy levels, or energy quanta, of the particle, which must always be greater than zero. Analogously, only certain harmonics are possible for a wave on a fixed length of string. Because only a limited set of energy levels are allowed by quantum theory, the particle is more likely to be in some places than in others.

More complicated systems have wavefunctions that are a combination of many sine waves and other mathematical functions, like a musical tone made up of many harmonics.

By bringing the wave–particle duality idea to atoms and all forms of matter, Schrödinger earned his place as one of the founding fathers of quantum mechanics.

> Quantum mechanics is certainly imposing. But an inner voice tells me that this is not yet the real thing. The theory says a lot, but does not bring us any closer to the secrets of the Old One. I, at any rate, am convinced that He is not playing at dice.
>
> Albert Einstein, letter to Max Born, 4 December 1926

The condensed idea
Harmonies of the atom

15 Heisenberg's uncertainty principle

I n 1927 Werner Heisenberg realized that some properties of the atomic world were inherently uncertain. If you know a particle's position then you cannot simultaneously know its momentum. If you know what time a particle did something, you cannot also pinpoint its energy.

In 1926, Werner Heisenberg and Erwin Schrödinger began an intense debate. Within a year of each other, the pair had presented radically different ways of expressing the quantized energy states of electrons in atoms, which had vastly different implications.

Heisenberg had proposed his 'matrix mechanics', a mathematical description of the ties between electron energy states and the spectral lines they produced as electrons performed quantum leaps between energy levels. It was a technical feat, but physicists were hesitant to take it up. They couldn't fathom what the equations – couched in unfamiliar matrix notation – really meant.

Boosted by the support of Albert Einstein, Schrödinger's alternative was much more palatable. Wave mechanics, which described electron energies in terms of standing waves or harmonics, involved familiar concepts. It sat easily with Louis de Broglie's suggestion that matter can behave like a wave, which was being confirmed through experiments showing that electrons can diffract and interfere.

In May 1926 Schrödinger published a paper proving that matrix and wave mechanics gave similar results – they were mathematically equivalent. He argued that his wave theory was better than the matrix description, which annoyed Heisenberg. One reason for Schrödinger's preference was that the discontinuities and quantum jumps that were intrinsic to matrix theory seemed unnatural. Continuous waves were much more pleasing. Heisenberg and Bohr thought those same jumps were the strength of their model.

Heisenberg was tetchy. He was a young man at a critical point in his career, actively trying to get a professorship in a German university. So he wasn't pleased that his great achievement was being overshadowed by the Austrian physicist.

Resolving the quantum standoff

In October 1926, Schrödinger came to Copenhagen to visit Niels Bohr. Heisenberg was also there, working with Bohr. The physicists argued face to face about the veracity of their ideas, but could not find a way forward. They went away to ponder the physical interpretation of their equations. Soon after, Heisenberg's colleague Pascual Jordan in Göttingen and Paul Dirac in Cambridge combined the equations of both pictures into one set of equations – the basis for what is now called quantum mechanics.

Physicists set about trying to explain what these equations meant in reality. How were 'classical' measurements made in the laboratory connected to what was going on at the scale of an atom?

Uncertainty the only certainty

While studying these equations, Heisenberg found a fundamental problem. He realized that it was impossible to measure some properties accurately because the apparatus used would interfere with the atoms being measured. A particle's position and momentum could not both be inferred at once; nor could its energy be known at a precise time. The reason was not the experimenter's lack of skill. These uncertainties lay at the heart of

> The more precisely the position is determined, the less precisely the momentum is known in this instant, and vice versa.
> Werner Heisenberg, 1927

quantum mechanics. Heisenberg presented his 'uncertainty principle' first in a letter to Wolfgang Pauli in February 1927, and later in a formal paper.

Any measurement has some degree of uncertainty associated with it. You might measure a child's height to be 1.20 metres, but your answer is only as good as the accuracy of your tape measure, say to a millimetre. Plus you could easily be off by a centimetre if the tape isn't taut or your eye doesn't line up with the child's head.

But Heisenberg's uncertainty isn't measurement error in this sense. His claim is profoundly different: you can never know both momentum and position at exactly the same time, no matter how precise an instrument you use. Should you pin one down, the other becomes more uncertain.

A trial by thought

Heisenberg imagined performing an experiment to measure the motion of a subatomic particle, such as a neutron. A radar could track the particle, by bouncing electromagnetic waves off it. For maximum accuracy you would choose gamma rays, which have very small wavelengths. However, because of wave–particle duality the gamma-ray beam hitting the neutron would act like a series of photon bullets. Gamma rays have very high frequencies and so each photon would carry a great deal of energy. As a hefty photon hit the neutron, it would give it a big kick that would alter its speed. So, even if you knew the position of the neutron at that instant, its speed would have been changed unpredictably.

If you used softer photons with lower energies, to minimize the velocity change, then their wavelengths would be longer and so the accuracy with which you could measure the position would now be degraded. No matter how you optimize the experiment, you cannot learn both the particle's position and speed simultaneously. There is a fundamental limit to what you can know about an atomic system.

Heisenberg realized the implications of his uncertainty principle were profound. Imagine a moving particle. Due to the fundamental limits on what you can know, you cannot describe the particle's past behaviour until a measurement ties it down. In Heisenberg's words, 'the path comes into existence only when we observe it'. Nor can the particle's future path be predicted, as you don't know its speed and position. Both the past and the future become blurred.

Newton overtaken

Such an unpredictable world clashed with physicists' interpretation of reality. Rather than the universe being filled with concrete entities – that exist independently and whose motions and properties could be verified through experiments – quantum mechanics revealed a seething mass of probabilities brought to fruition only by the action of an observer.

There is no cause and effect, only chance. Many physicists found this difficult to accept – Einstein never did. But that is what the experiments and the mathematics tell us. Physics stepped beyond the laboratory of experience into the abstract realm.

Werner Heisenberg (1901–76)

Werner Heisenberg grew up in Munich, Germany, and loved the mountains. As a teenager during the First World War he worked on a dairy farm, reading mathematics and playing chess in his spare time. At Munich University he studied theoretical physics, completing a doctorate unusually early. Heisenberg took up a professorship at Leipzig aged just 25, and worked in Munich, Göttingen, and Copenhagen, where he met Niels Bohr and Albert Einstein. In 1925 he invented matrix mechanics, receiving the Nobel Prize in 1932. His uncertainty principle was formulated in 1927. During the Second World War Heisenberg led the German nuclear weapons project, which was ultimately unsuccessful in producing a bomb. Whether he deliberately delayed the project or lacked the resources is still debated.

The condensed idea
Known unknowns

16 The Copenhagen interpretation

I n 1927 the Danish physicist Niels Bohr tried to explain the physical meaning of quantum mechanics. In what became known as the Copenhagen interpretation, he combined Heisenberg's uncertainty principle and Schrödinger's wave equation to explain how an observer's intervention means that there are things we can never know.

The quest to understand the meaning of quantum mechanics began in earnest in 1927. Physicists fell into two camps. Werner Heisenberg and his colleagues believed that the particle nature of electromagnetic waves and matter, described in his matrix representation, was paramount. Erwin Schrödinger's followers argued that the physics of waves underlay quantum behaviour.

Heisenberg had also shown that our understanding was fundamentally limited by his uncertainty principle. He believed that both the past and future were unknowable until fixed by observation because of the intrinsic uncertainty of all the parameters describing a subatomic particle's movement.

Another man tried to pull everything together. Bohr, the head of Heisenberg's department at the University of Copenhagen, was the scientist who had a decade earlier explained the quantum energy states of electrons in the hydrogen atom. When Heisenberg came up with his 'uncertainty principle' in 1927 he was working in Copenhagen at Bohr's institute. Bohr apparently returned from a skiing trip to find Heisenberg's draft paper on his desk, and a request to forward the document to Albert Einstein.

> **Anyone who is not shocked by quantum theory has not understood it.**
> Niels Bohr, 1958

Bohr was intrigued by the idea, but complained to Einstein that Heisenberg's imagined test – involving a gamma-ray microscope – was flawed as it did not consider the wave properties of matter. Heisenberg added a correction that included the scattering of light waves, but his conclusion still held firm. Uncertainties were inherent in quantum mechanics. But what was really going on?

A coin forever spinning

In Bohr's view, the wave and particle aspects of a real entity were 'complementary' characteristics. They are two sides of the same coin, in the same way that some illusions trick our eyes into seeing two different pictures in a black and white pattern – a vase or two profiles facing one another, say.

The real electron, proton or neutron is neither one nor the other, but a composite of both. A given trait only appears when an experimenter intervenes and selects which aspect to measure. Light appears to behave like a photon or electromagnetic wave because that is the sign we are looking for. Because the experimenter disturbs the pristine system, Bohr argued, there are limits to what we can know about nature. The act of observation generates the uncertainties that Heisenberg spotted. This line of reasoning became known as the 'Copenhagen interpretation' of quantum mechanics.

The uncertainty principle, which states that one cannot measure both the position and momentum of any subatomic particle at the same time, is central, Bohr realized. Once one characteristic is measured precisely, the other becomes less well known. Heisenberg believed that the uncertainty arose due to the mechanics of the measurement process itself. To measure a quantity we must interact with it, such as by bouncing photons off a particle to detect its movement. That interaction changes the system, Heisenberg realized, making its subsequent state uncertain.

The undetached observer

Bohr's understanding was quite different: the observer is part of the system being measured, he argued. It doesn't make sense to describe the subject without including the measuring device. How can we describe a particle's motion by considering it alone if it is being bombarded with photons in order to track it? Even the word 'observer' is wrong, said Bohr, because it suggests an external entity. The act of observation is like a switch, which determines the system's final state. Before that point we can only say that the system has some chance of being in some possible state.

When Bohr is about everything is somehow different. Even the dullest gets a fit of brilliance.
Isidor I. Rabi in Daniel J. Kevles, *The Physicists* (1978)

What happens when we make a measurement? Why does light passing through two slits interfere like waves one day, but switch to particle-like behaviour the next if we try to catch the photon as it passes through one slit? According to Bohr, we choose in advance how it turns out by deciding how we would like to measure it.

Correspondence principle

To bridge the gap between quantum and normal systems, including our own experiences on human scales, Bohr also introduced the 'correspondence principle', that quantum behaviour must disappear for larger systems that we are familiar with, when Newtonian physics is adequate.

What we can know

Here Bohr turned to Schrödinger's equation and his concept of the 'wavefunction', containing everything we can know about a particle. When an object's character is fixed, say as a particle or a wave, by an act of observation we say that the wavefunction has 'collapsed'. All the probabilities, bar one, vanish. Only the outcome lingers. So a beam of light's wavefunction is a blend of two possibilities: whether it behaves as a wave or a particle. When we detect the light, the wavefunction collapses to leave one form, not because it switches its behaviour, but because light truly is both.

Heisenberg initially rejected Bohr's picture. He clung to his original imagery of particles and energy jumps. The two fell out. Heisenberg apparently burst into tears at one point during an argument with Bohr. The stakes for the young man's career were high.

Matters improved later in 1927 when Heisenberg secured a job at the University of Leipzig. Bohr presented his complementarity idea to great acclaim at a conference in Italy and many physicists took it up. By October, Heisenberg and Max Born were talking of quantum mechanics as having been fully solved.

Not everyone agreed, notably Einstein and Schrödinger, who remained unconvinced by Bohr's doctrine for the rest of their days. Einstein believed that particles could be measured precisely. The idea that real particles were governed by probabilities unsettled him. These would not be needed in a better theory, he argued. Quantum mechanics must be incomplete.

Even today physicists struggle to comprehend the deep meaning of quantum mechanics. Some have tried to offer new explanations, although none has overturned Bohr's. The Copenhagen view has stood the test of time because of its explanatory power.

The condensed idea
Some things we may never know

17 Schrödinger's cat

To reveal how ridiculous the Copenhagen interpretation of quantum mechanics was, Erwin Schrödinger pointedly used a cat as a case study. Imagining it boxed in for a period with a vial of poison, he argued that it made no sense to think of a real animal as a probability cloud simply because we lack knowledge about what is going on.

Niels Bohr's 1927 proposal of the Copenhagen interpretation of quantum mechanics thrilled many physicists, but the hard core fans of the wavefunction approach did not get on board. Erwin Schrödinger and Albert Einstein stayed on the sidelines.

In 1935, Schrödinger tried to ridicule Bohr's idea of a fuzzy probabilistic quantum world by publishing a hypothetical situation that illustrated the counter-intuitive nature of collapsing wavefunctions and observer influence. Albert Einstein did likewise, through his Einstein–Podolsky–Rosen paradox paper, which hinted at implausible long-distance correlations.

In the Copenhagen interpretation, quantum systems are dark and indeterminate until an observer steps in, flicking the light switch and deciding what quality his or her experiment will measure. Light is both particle and wave until we decide which form we want to test – then it adopts that form.

Schrödinger, who was keen on developing a wave-based theory of atoms, disliked the idea that something unseen 'existed' in all possible forms. When you open a fridge to see that it contains cheese, celery and milk, was it really in a mathematical quandary about whether to reveal chocolate and an egg before you looked?

Quantum probabilities obviously made little sense on the grand scale. Schrödinger's article contained a thought experiment that tried to illustrate this behaviour using a more emotionally engaging subject – a cat.

Quantum limbo

Schrödinger considered the following scenario. A cat is locked in a steel chamber, along with a 'diabolical device': a flask of poisonous hydrocyanic acid, to be shattered only upon the decay of a radioactive atom. The cat's fate depends on the probability of whether the atom has decayed or not.

'If one has left this entire system to itself for an hour, one would say that the cat still lives if meanwhile no atom has decayed. The first atomic decay would have poisoned it,' he wrote. Schrödinger's depressing apparatus would deliver a 50:50 chance of the cat being either alive or dead when the box is opened after that time.

According to the Copenhagen interpretation of quantum physics, while the box is closed, the cat exists in a blend of states – being both alive and dead at the same time. Only when the box is opened is the animal's fate sealed, just as a photon is both a wave and a particle until we choose how to detect it, when its wavefunction collapses to favour one facet.

Schrödinger argued that such an abstract explanation made no sense for a real animal like a cat. Surely it was either alive or dead, not a mixture of both. Bohr's interpretation, he reasoned, must be convenient shorthand for what was really going on at a deeper level. The universe operates in unseen ways, and we can only witness part of the picture at any one time.

Einstein also thought the Copenhagen picture was nonsensical. It raised many more questions. How does the act of observation cause the wavefunction to collapse? Who or what can do the observing – does it have to be a human or can any sentient being do so? Can the cat observe itself? Is consciousness necessary? Can the cat collapse the decaying particle's wavefunction to dictate that outcome? How does anything in the universe exist, for that matter? Who observed the first star, say, or the first galaxy? Or were they in a quantum quandary until life got going? The riddles are endless.

I am convinced that theoretical physics is actual philosophy.
Max Born, *My Life and My Views* (1968)

Following the Copenhagen logic to its extreme, it's possible that nothing in the universe exists as such. This view is reminiscent of the philosophy of George Berkeley, a 17th-century philosopher and contemporary of Isaac Newton. Berkeley presented the case that the entire external world is only a part of our imagination. We can have no evidence that anything external to ourselves exists – all that we can sense or know is contained within our mind.

Many worlds

The issue of how measurements cement outcomes was revisited in a novel way in 1957 by Hugh Everett. He suggested that observations don't destroy options but shear them off into a set of parallel universes.

According to his 'many worlds' hypothesis, every time we pin down a photon's character, the universe splits in two. In one world light is a wave; in the other it is a particle. In one universe the cat is alive when we open the box; in the complementary dimension the animal has been killed by the radioactive poison.

In all other respects both forks of the universe are the same. So every observation produces a new world, building up a series of branches. Over the history of the universe that could make for a lot of parallel worlds – an indefinite, perhaps an infinite, number.

Everett's idea was ignored at first, until a popular physics article and science fiction fans, struck by its appeal, brought it into the limelight. But it now chimes with a modern variant called the 'multiverse' theory, which some physicists are using to explain why the universe is so hospitable – because all the inhospitable universes are hived off elsewhere.

Einstein argued that there should exist something like a real world, not necessarily represented by a wave function, whereas Bohr stressed that the wave function doesn't describe a 'real' microworld but only 'knowledge' that is useful for making predictions.
Sir Roger Penrose, 1994

Erwin Schrödinger (1887–1961)

Erwin Schrödinger was born in Vienna, the son of a botanist. He chose to study theoretical physics at university, although he was also interested in poetry and philosophy. During the First World War, he served with the Austrian artillery in Italy, keeping up his physics research while on the front.

Schrödinger returned to academic posts in universities including Zurich and Berlin. But as the Nazis came to power, he decided to leave Germany and moved to Oxford. Soon after he arrived there he heard he had won the 1933 Nobel Prize, with Paul Dirac, for quantum mechanics. In 1936 he returned to Graz in Austria, but political events again overtook him. He lost his job after criticizing the Nazis, and eventually moved to the Institute for Advanced Studies in Dublin, where he stayed until retiring to Vienna. Schrödinger's personal life was complicated: he had numerous affairs, many with his wife's knowledge, and had several children with other women.

The condensed idea
Dead and alive

18 The EPR paradox

I n 1935 three physicists – Albert Einstein, Boris Podolsky and Nathan Rosen – came up with a paradox that challenged quantum-mechanical interpretations. The fact that quantum information would apparently have to travel faster than the speed of light seemed to poke holes in the idea of collapsing wavefunctions.

The Copenhagen interpretation of quantum mechanics, proposed by Niels Bohr in 1927, reasons that the act of measurement influences a quantum system, causing it to adopt the characteristics that are subsequently observed. Light's wave- and particle-like properties know when to appear because the experimenter effectively tells them what to do.

Albert Einstein thought this preposterous. Bohr's idea meant that quantum systems remained in limbo until they were actually observed. Before some measurement tells you what state it is in, the system exists in a blend of all possible states of being. Einstein argued that such a superposition was unrealistic. A particle exists whether or not we are there to see it.

Einstein believed that everything in the universe exists in its own right, and the uncertainties of quantum mechanics illustrated that something was wrong with the theory and our interpretation of it. To expose gaps in the Copenhagen view, Einstein, together with his colleagues Boris Podolsky and Nathan Rosen, came up with an imaginary experiment, which they published in a paper in 1935. This is known as the Einstein–Podolsky–Rosen, or EPR, paradox.

Imagine a particle, perhaps an atomic nucleus, which decays into two smaller ones. According to energy-conservation rules, if the mother particle was originally stationary, the daughter particles must have equal and opposite linear and angular momentum. The emerging particles fly apart and spin in opposite directions.

Other quantum properties of the pair are also linked. If we measure the spin direction of one particle, we instantly know the state of the other: it must have the opposite spin to conform to quantum rules. As long as neither particle interacts with others, which would scramble the signal, this fact remains true no matter how far apart the particles get or how much time elapses.

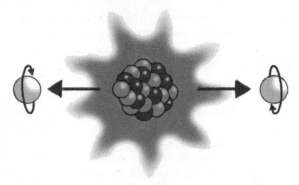

An atomic nucleus decays creating two particles of opposite spin.

In the language of the Copenhagen interpretation, both daughter particles exist at first in a superposition of all possible outcomes – a mixture of all the different speeds and spinning directions that they could have. At the moment that we make a measurement of one, the wavefunction probabilities for both particles collapse down to consolidate that result. Einstein, Podolsky and Rosen argued that this made no sense. Einstein knew that nothing can travel faster than light. So how could you pass an instant signal to a particle that could be very far away, potentially on the other side of the universe? The Copenhagen interpretation must be wrong. Schrödinger later used the term 'entanglement' to describe this weird action at a distance.

> Quantum theory thus reveals a basic oneness of the universe.
>
> Fritjof Capra, *The Tao of Physics* (1975)

Entanglement

Einstein believed in 'local reality', that everything in the world exists independently of us and signals carry information no faster than the speed of light. The two particles in the thought experiment must already know which state each is in when they separate, he reasoned. They carry that knowledge with them, rather than switching state simultaneously at remote distances.

But Einstein was wrong. His idea sounds reasonable and chimes with our everyday experience. But it has been demonstrated to be false by numerous quantum experiments over many decades. The 'spooky action at a distance' does take place, and coupled particles do seem to 'speak' to one another across space faster than light. Physicists have now entangled the quantum properties of more than two particles and have seen them switch states together across many tens of kilometres.

Quantum signalling at a distance opens up a host of applications for new forms of remote communications, including sending instant messages across vast reaches of space. It raises possibilities for quantum computers, able to carry out many calculations at the same time across the entire memory of the machine.

The units of quantum information are known as 'quantum bits', or 'qubits' for short. Just as normal computers use binary code to describe messages as long sentences of 0s and 1s, qubits would adopt one of two quantum states. But better than that, they could also exist in a blend of states, thus allowing calculations to be performed that we can only dream of.

> As far as the laws of mathematics refer to reality, they are not certain; and as far as they are certain, they do not refer to reality.
>
> Albert Einstein,
> Sideslights on Relativity (1920)

Yet the indeterminacy that gives quantum signalling its power also means that we cannot transmit a complete set of information from one place to another. Heisenberg's uncertainty principle always means there's a gap somewhere in what we can know. So human teleportation – as we know it in science fiction – is impossible.

Action at a distance

While transmitting atoms will never happen, it is possible to shift information across space using quantum teleportation. If two people – often called Alice and Bob by physicists – each hold one of a pair of entangled particles, through making particular measurements they can use them to convey qubits. First, Alice and Bob need to acquire their pair of coupled particles, perhaps two photons, each taking one away. Alice's qubit may be in some state that she wants to send to Bob. Even if she doesn't know what that state is, she can influence Bob's photon to give him that message. By making a measurement on her photon, Alice destroys it. But Bob's photon takes it on. Bob can make his own measurement to extract the information.

Because nothing actually travels anywhere, there is no teleportation of matter in that sense. Apart from the first exchange of the particles, there is no direct communication between the two messengers. Rather, Alice's original message is destroyed in the sending process and its content is recreated somewhere else. Entangled particles can also be used to transmit coded messages, so that only the intended receiver can read them. Any eavesdropper would break the purity of the entanglement, ruining the message for good.

Einstein's discomfort with entanglement was understandable – it is difficult to imagine the universe as a web of quantum connections, with unknown numbers of particles talking to their distant twins. But that is how it is. The universe is one big quantum system.

The condensed idea
Instant messaging

19 Quantum tunnelling

Radioactivity can be explained only with quantum mechanics. An alpha particle may need a lot of energy to escape the strong glue of the nucleus, but because there is a small probability that it can do so, there's a chance that the particle will exceed the energy barrier. This is called quantum tunnelling. When you throw a tennis ball against a wall you expect it to bounce back. Imagine if it appeared instead on the other side of the wall. This is what can happen at the atomic scale according to the rules of quantum mechanics.

Because a particle, a molecule, or even a cat, may be described as a wave – as embodied in the wavefunction of the Schrödinger equation – there is some chance that it is extensive. Electrons, for example, do not orbit their nuclei like planets, but are spread over their entire orbital shells. If we think of it as a particle, the electron may be anywhere within that region, with some probability. It is unlikely, but electrons may even jump out of their host atoms.

Quantum tunnelling is the ability of a particle to achieve an energy feat in the quantum world that it could not in the classical picture. It is as if a horse could somehow get to the other side of a hedge too high to jump because its wavefunction can burrow through it. Overcoming energy barriers through tunnelling plays a role in the nuclear fusion processes that light up our Sun and other stars, and has applications in electronics and optics.

Radioactive decay

Physicists came up with the idea of quantum tunnelling when trying to work out how radioactive atoms decay. It is impossible to predict the exact moment when an unstable nucleus will break apart and kick out a lump of radiation, but on average for many nuclei we can say how likely it is. This information is usually given in terms of the 'half-life', the length of time it will take for around half of the atoms to decay. More formally, it is the time by which there's a 50% chance that an atom will have decayed.

In 1926 Friedrich Hund came up with the concept of quantum tunnelling, and it was soon invoked to explain alpha decay. A chunk of polonium-212, for example, readily emits alpha particles (two protons and two neutrons) with a half-life of 0.3 microseconds. These have typical energies of around 9 MeV (million electronvolts). But the alpha particle should require 26 MeV to escape the binding energy of the nucleus, according to classical physics. It shouldn't be able to jump out at all, but clearly it does. What is going on?

Because of the uncertainties of quantum mechanics, there is a small possibility that an alpha particle will escape the polonium atom. The alpha particle is able to leap – or quantum-tunnel – across the high-energy barrier. The probability that it does so can be calculated using Schrödinger's wave equation, by extending the wavefunction beyond the atom. Max Born realized that tunnelling was a general feature of quantum mechanics and not restricted to nuclear physics.

How can we envisage quantum tunnelling? The alpha particle feeling the attractive pull of the nuclear force is like a ball rolling in a valley. If it has a small amount of energy it rolls back and forth but is trapped. If it gains enough energy it could roll over the hill and escape the valley. This is the classical physics picture. In the quantum world, the alpha particle also has wave-like tendencies. And these can spread out. According to Schrödinger's wave equation, the particle's properties can be described by a wavefunction, which looks roughly like a sine wave. The wavefunction must be continuous, and reflect the fact that the particle is most likely to exist within the atom, but there's also a small probability that the particle could escape the valley of the nuclear charge, so a little must leak out.

Friedrich Hund (1896–1997)

Hund grew up in the German town of Karlsruhe. He studied mathematics, physics and geography in Marburg and Göttingen, eventually returning to a position at Göttingen in 1957. Hund visited Niels Bohr in Copenhagen and was a colleague of Erwin Schrödinger and Werner Heisenberg. He worked with Max Born on a quantum interpretation of the spectra of diatomic molecules, such as molecular hydrogen. In 1926 he discovered quantum tunnelling. Hund's rules for filling up electron shells are still widely used in physics and chemistry.

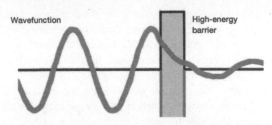

Wavefunction

High-energy barrier

There is a small chance a particle's wavefunction can 'tunnel' through an energy barrier, even though it does not have enough energy according to classical physics to overcome it.

Visualizing it mathematically, the wavefunction is a sine wave within the valley, but when it reaches the sides of the hill it extends right through that energy barrier. It drops off steadily in strength as it travels through, so a thicker barrier is harder but not impossible to penetrate. And then it resumes its wiggly sine-wave character on the other side of the hill. By calculating the strength of the wavefunction on the far side of the hill relative to that inside, the probability that the alpha particle escapes can be worked out.

Evanescent waves

Light can spread energy through a mirror thanks to a related phenomenon. A beam of light that skims a mirror and is completely reflected cannot be explained simply using Maxwell's equations of electromagnetic waves. To keep the waves' properties unbroken and balance the equations, a little energy must pass through the mirror. This is known as an evanescent wave.

> Elementary particles and the atoms they form are doing a million seemingly impossible things at once.
> Lawrence M. Krauss, 2012

Evanescent waves decay exponentially in strength, and quickly become so weak that they are invisible. But if another equivalent material is placed near the first mirror the energy can be picked up and transmitted. This coupling technique is used for some optical devices and is analogous to the spread of magnetic energy between inducting coils in a transformer.

Tunnelling is also useful in electronics. It allows electrons to jump barriers in a controlled way in arrays of semiconductors and superconductors. Tunnel junctions, for example, are made from two conducting materials with an insulator sandwiched in between – a few electrons can jump from one side to the other through the insulator. Tunnelling is also used in some types of diode and transistor, as a means of controlling voltages, a bit like a volume control.

The scanning tunnelling microscope uses the principle to image surfaces of materials, revealing details on the scale of atoms. It does so by placing a charged needle close to the surface. A small number of electrons quantum-tunnel off the needle towards the surface, and the strength of the current gives away the distance between them. Such microscopes are so powerful they are accurate to within 1% of an atom's diameter.

> With the advent of quantum mechanics, the clockwork world has become a lottery. Fundamental events, such as the decay of a radioactive atom, are held to be determined by chance, not law.
>
> Ian Stewart, *Does God Play Dice?* (2002)

The condensed idea
Shortcut through the hill

20 Nuclear fission

Once the neutron was discovered, physicists started to fire them at large atoms, expecting to build new isotopes and elements. Instead the nucleus split apart – it underwent fission. Energy was released in the process, making fission a new goal as a power source and for the atom bomb.

In the 1920s and 30s physicists looked beyond electrons towards probing the atomic nucleus. Radioactivity – in which a large nucleus such as uranium or polonium breaks apart to shed smaller constituents – was well known. But the means by which it did so were unclear.

Following on from his 1911 discovery of the nucleus in his gold-foil experiment, Ernest Rutherford transmuted nitrogen into oxygen by firing alpha particles at the gas in 1917. Physicists knocked small bits off other nuclei. But it was not until 1932 that John Cockcroft and Ernest Walton in Cambridge split an atom in half by firing fast protons at a lithium target. The same year the opposite experiment – gluing two nuclei together in nuclear fusion – succeeded, when Mark Oliphant fused two deuterium (a heavy form of hydrogen) nuclei to make helium.

James Chadwick's discovery of the neutron, also in 1932, opened up new possibilities. Enrico Fermi in Italy and Otto Hahn and Fritz Strassmann in Germany shot neutrons into the heavy element uranium, attempting to create even heavier atoms. But in 1938, the German pair did something more profound. They sheared a massive uranium nucleus roughly in half, giving off barium, which is 40% as massive.

For something less than half a per cent of the mass of the target atom, the neutron's impact on the uranium seemed excessive. It was as if a melon split in two when hit by a pea. The finding was also unexpected because physicists at the time, including George Gamow and Niels Bohr, thought that the nucleus was like a liquid drop. Forces of surface tension should resist the droplet's division, and even if it did split then the two positively charged drops of substance would repel and fly apart, they believed. That wasn't what was seen.

The solution came from Hahn's colleague Lise Meitner. Exiled in Sweden, having fled Nazi Germany, Meitner and her physicist nephew

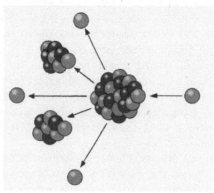
Neutrons fired at a heavy nucleus can split it in half.

Otto Frisch quickly realized that it wasn't so odd for a large nucleus to shear in half – each product would be more stable than the original and so the total energy would be less overall. The remaining energy would be given off. Meitner and Frisch named the process 'fission' after the division of a biological cell.

A potential weapon

On his return to Denmark, Frisch mentioned their idea to Niels Bohr, who spread it over the Atlantic during his lecture tour. At Columbia University in New York, the Italian émigré Enrico Fermi began fission experiments in the basement. The Hungarian exile Léo Szilárd, also in the US, realized that this uranium reaction could produce spare neutrons that would produce more fission – so causing a nuclear chain reaction (a self-sustaining reaction) that could yield a vast amount of explosive energy.

The Second World War had broken out, and Szilárd worried that German scientists would come to the same conclusions. He and Fermi agreed not to publish their finding. In 1939 Szilárd, with two other Hungarian refugees, Edward Teller and Eugene Wigner, persuaded Albert Einstein to add his name to a letter warning US President Franklin Roosevelt of the risk that such a reaction could be used to build an atomic bomb.

Frisch, now exiled in England, also set to work with Rudolph Peierls to work out how much uranium was needed and of what sort.

His answer was shocking – mere kilograms of a uranium isotope with atomic weight 235 (^{235}U) would be enough to produce a chain reaction, not the tonnes of material first suspected.

Ideas were again shared across the Atlantic, but it still proved difficult to ignite a chain reaction in the laboratory. Purifying uranium was hard, and neutrons in the experiments were quickly mopped up before they could trigger the fission cascade. Fermi obtained the first chain reaction in 1942 at the University of Chicago, beneath the football stadium.

Meanwhile, in Germany, Werner Heisenberg had also flagged up to the government the possibility of a uranium-based bomb. Fortunately for the outcome of the war, the German effort fell behind that of the Allies. Heisenberg's position in this is unclear – some people think that he deliberately dragged his feet, others have tarred him for taking a leading role in the programme. Nevertheless, despite German scientists having discovered fission, by the war's end the country had not even achieved a chain reaction.

Robert Oppenheimer (1904–67)

Robert Oppenheimer was born in New York City into a wealthy family. He first went to New Mexico whilst a teenager, sent there to recover from an illness. At Harvard University he studied chemistry and physics, moving to Cambridge in 1924. Oppenheimer didn't get on with his tutor, Patrick Blackett, and claimed to have left an apple covered with chemicals on his desk.

In 1926 Oppenheimer moved to Göttingen to work with Max Born, where he also encountered giants such as Heisenberg, Pauli and Fermi. Oppenheimer returned to the USA in the 1930s and worked at Caltech and Berkeley. Described as both mesmerizing and frigid, he had a powerful personality. His communist leaning led to him being distrusted by the government. Yet in 1942 they asked him to lead the Manhattan Project. Oppenheimer was troubled by the dropping of the atomic bomb, quoting from the Bhagavad Gita: 'Now I am become Death, the destroyer of worlds.' In his later life he joined with other physicists to promote global nuclear peace.

In September 1941, Heisenberg visited German-occupied Copenhagen and looked up his old colleague Bohr. The subject of their conversation is not known exactly – it is the theme of Michael Frayn's play *Copenhagen* – although both gave later accounts in letters, some never sent. Bohr's letters have only recently been made public by his family. One remarks that Heisenberg told him in confidence about the German atomic war effort. Bohr was clearly upset and tried to send a message to London via Sweden. But the message was garbled and wasn't understood when it arrived.

> No-one really thought of fission before its discovery.
> Lise Meitner, 1963

The Manhattan Project

Back in the US, Frisch's realization that only a little uranium was needed to make a bomb coincided with the Japanese attack at Pearl Harbor. Roosevelt launched the US nuclear bomb project, known as the Manhattan Project. It was led by the Berkeley physicist Robert Oppenheimer from a secret base at Los Alamos in New Mexico.

Oppenheimer's team started designing the bomb in the summer of 1942. The trick was to keep the quantity of uranium below the critical mass until the detonation set the fission in progress. Two methods were tried, fashioned into bombs called 'Little Boy' and 'Fat Man'. On 6 August 1945, 'Little Boy' fell on the Japanese city of Hiroshima, releasing the equivalent of about 20,000 tonnes of dynamite. Three days later, 'Fat Man' delivered its blow to Nagasaki. Each bomb killed around 100,000 people instantly.

The condensed idea
Nuclear splitting

21 Antimatter

Most elementary particles have a mirror-image twin. Antimatter particles have the opposite charge but the same mass as their companion. A positron for example is a positively charged version of the electron. Most of the universe is made of matter. When matter and antimatter meet they annihilate in a puff of pure energy.

In 1928, the physicist Paul Dirac attempted to upgrade Schrödinger's wave equation by adding in the effects of special relativity. The wave equation described particles such as electrons in terms of the physics of standing waves, but at that time was incomplete.

The theory applied to particles with little energy or travelling slowly, but didn't explain the relativistic effects of energetic particles, such as the outer electrons in atoms larger than hydrogen. To better fit the spectra of large or excited states of atoms, Dirac worked in the relativistic effects – length contraction and time dilation – to show how they affected the shapes of the electron orbits.

While it worked in predicting the size of the electrons' energies, Dirac's equation at first seemed too general. The mathematics allowed the possibility that electrons could have negative as well as positive energies, just as the equation $x^2 = 4$ has the solutions $x = 2$ and $x = -2$. The positive-energy solution was expected, but negative energy made no sense.

Equal, opposite, but rare

Rather than ignore this confusing 'negative energy' term, Dirac suggested that such particles might actually exist. Perhaps there were forms of electrons with positive rather than negative charge but with the same mass? Or they might be thought of as 'holes' in a sea of normal electrons. This complementary state of matter is called 'anti'-matter.

The hunt began, and in 1932 the Caltech scientist Carl Anderson confirmed that positrons exist. He was following the tracks of showers of particles produced by cosmic rays – very energetic particles that crash into the atmosphere from space, first seen by the German scientist Victor Hess two decades earlier. Anderson saw the track of a positively charged particle with the electron's mass, the positron. Antimatter was no longer just an abstract idea but was real.

The next antiparticle, the antiproton, was detected two decades later, in 1955. Emilio Segrè and his team, working at the University of California at Berkeley, used a particle accelerator – a machine called a bevatron – to fire a stream of fast protons at nuclei in a fixed target. The energies of the protons were high enough that antiparticles were produced in the collsions. A year later, the antineutron was also found.

With the antimatter equivalent building blocks in place, was it possible to build an anti-atom, or at least an anti-nucleus? The answer, shown in 1965, was yes. A heavy hydrogen (deuterium) anti-nucleus (an anti-deuteron), containing an antiproton and antineutron, was created by scientists at CERN in Europe and Brookhaven Laboratory in America. Tagging on a positron to an antiproton to make a hydrogen anti-atom (anti-hydrogen) took a little longer, but was achieved also at CERN in 1995. Today experimenters are testing whether anti-hydrogen behaves in the same way as normal hydrogen.

To deliberately create antimatter on Earth – rather than catching its signs in cosmic rays from space – physicists need special machines that use huge magnets to boost the speeds of particles and focus them into beams. In giant particle accelerators, such as those at CERN in

Paul Dirac (1902–84)

Paul Dirac has been called 'the strangest man'. He admitted that he never started a sentence without knowing how to finish it; people joked that his only words were 'Yes', 'No', and 'I don't know'. Luckily he was as brilliantly clever as he was acutely shy. His doctorate, completed at Cambridge University in record time and characteristic brevity, was a completely new picture of quantum mechanics. Dirac went on to incorporate relativity theory into quantum theory and to predict the existence of antimatter, plus pioneering work in early quantum field theory. When he was awarded the Nobel Prize in 1933, fearing the publicity, Dirac hesitated to go and accept it. He only agreed when he was told that he might get more attention if he turned it down.

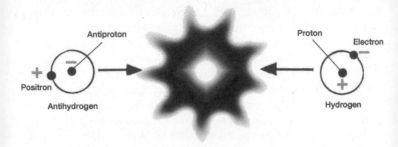

Matter and antimatter annihilate to form pure energy.

Switzerland or Fermilab near Chicago, streams of particles can be fired at targets or at other beams, releasing energy according to Einstein's $E = mc^2$ equation that then creates a shower of other particles. Because matter and antimatter annihilate to give a flash of pure energy, if you meet your antimatter twin, think twice before shaking hands.

The Bang's imbalance

When we look out into the universe, we don't see many flashes of annihilating particles. The reason is that it is almost entirely made of matter – less than 0.01% of the stuff in the universe is made of antimatter. What caused this fundamental imbalance?

> I like to play about with equations, just looking for beautiful mathematical relations which maybe don't have any physical meaning at all. Sometimes they do.
> Paul Dirac, 1963

It could be that slightly dissimilar amounts of each were created in the Big Bang. Over time most of the particles and antiparticles have collided and mopped each other up but a few remain. If only 1 in every 10,000,000,000 (10^{10}) matter particles survived, it would explain the proportions we see now. This could also explain the large numbers of photons and raw forms of energy that pepper the universe.

Or it could be that some quantum process in the very early universe set a switch to favour matter over its mirror form. Perhaps some unusual particles were created in the fireball that decayed predominantly to matter. Whatever the reason, thousands of physicists at the world's great particle accelerators are trying to find it out.

I think that the discovery of antimatter was perhaps the biggest jump of all the big jumps in physics in our century.
Werner Heisenberg, quoted in 1973

The condensed idea
Equal and opposite

22 Quantum field theory

I f light and electromagnetic waves can be carried by photons, then quantum field theory supposes that all fields are carried across space by fundamental particles. Also implied is that particles of any one type are indistinguishable, particles are emitted and absorbed during interactions, and antimatter exists.

Hold two magnets close to one another, you can feel them repel. But how is that force being transmitted? Likewise, how does the Sun's light or its gravity manage to stretch across vast expanses of space to influence Earth, or tiny Pluto?

The idea that forces act across distance through some extended 'field' grew out of Michael Faraday's work on electricity and magnetism in the mid-19th century. His search for fundamental laws of electromagnetism – linking all electric and magnetic phenomena – was completed decades later by James Clerk Maxwell. In just four equations, Maxwell described the various aspects of electromagnetic fields, including how they fall off in strength with distance.

But how are forces communicated? In the classical world of physics, we usually think of objects as taking energy from one place to another. In a gun, atoms from a pressure wave transfer the energy of an explosion to a bullet, which punches a target. In the early 20th century, Albert Einstein described light in a similar way, as a stream of photons depositing packets of energy when they hit a metal plate. But what about the other forces: gravity, or the weak and strong forces that glue the nucleus together?

Force-carrying particles

Quantum field theory, which emerged in the 1920s, supposes that all fields transmit their energy through flows of quantum particles – known as 'gauge bosons'. Like photons, they traverse space to deliver their blow. Like photons, they carry certain 'quanta' of energy. But unlike photons, some of the force-field carriers have mass. And there is a whole menagerie of these.

Force-carrying particles are not like hard billiard balls, but are ripples in the underlying energy field. They are neither truly waves nor particles, but something in between, as the quantum pioneers

Niels Bohr and Louis de Broglie explained was true of everything at the atomic scale. The force carriers – including and similar to photons – can act like particles in circumstances that demand it, and they can only carry certain amounts of energy, according to the quantum rules. Fermions, like the electron, can also be thought of as carriers of their associated fields.

Dirac and quantum theory

The first field where quantum behaviour was studied was the electromagnetic field. In the 1920s the British physicist Paul Dirac tried to develop a quantum theory of electromagnetism, which he published in 1927. His focus was the electron. What made it tricky to describe its behaviour was that he needed to explain how a photon could be emitted when an electron drops from a high- to a lower-energy orbital in an atom. How was that second particle actually created?

He reasoned that, just as chemicals interact, so can particles interact, as long as they follow quantum rules. Certain quantities – such as charge and energy – must be conserved before and after the interaction, if you consider all the particles. So an electron undergoes an interaction as it drops in energy, emitting the energy difference in the form of a photon.

Dirac's wrangling with his equations for electrons eventually led to his prediction of antimatter and the positron – which he visualized as a hole in a sea of electrons. Particles have antiparticle twins, with opposite charge and negative energy. The positron is the anti-electron.

An assumption of quantum field theory is that all these elementary particles are indistinguishable. One photon with a particular energy looks and behaves much like any other, no matter where it is in the universe; all electrons are pretty much the same, whether they are in a piece of sulphur, a copper sheet, or whizzing through a neon gas tube.

Birth and death of energy

Particles can sometimes pop into and out of existence. In accordance with Heisenberg's uncertainty principle, there is a small chance that a packet of energy may spontaneously appear for a while, even in the vacuum of space. The chance of it doing so is linked to the product of the particle's energy and the time for which it appears – an energetic particle can only pop into existence for a short time.

Dealing with this eventuality meant that quantum field theory had to handle the statistics of many particles, and include the Pauli exclusion principle that states that no two fermions can have the same properties. Pascual Jordan and Eugene Wigner in 1927 and 1928 worked out how to statistically combine many wavefunctions to represent fields.

But early quantum field theories struggled to explain some phenomena. One was the fact that the fields produced by the force carriers affected the particles themselves. For example, an electron has an electric charge, so it produces and sits within its own field. Inside an atom, this causes the electron orbital energies to shift a little.

The very idea of what an electron or photon was made of was difficult to visualize. If the negatively charged electron was extended, and not a point source, some bits would repel others. Electromagnetic stresses could rip it apart. But if electrons had no extent, then how do you assign properties like charge and mass to an infinitely small point? The equations quickly filled with infinities.

In 1947, physicists found a way of cancelling out the infinities – known as renormalization – and pioneers like Julian Schwinger and

Richard Feynman pushed the theory further. The result, known as quantum electrodynamics (QED), described how light and matter interact and was consistent with relativity. Electromagnetic effects were transmitted across space by the massless photon over large distances.

Explaining the other forces was harder and took decades. The unification of the electromagnetic and the weak nuclear force – which is involved with fusion and beta radioactive decay – awaited a better understanding of protons and neutrons, which are built from tiny quarks. The strong nuclear force was an even greater challenge, owing to the tiny range over which it operates. So electroweak theory and quantum chromodynamics were only developed in the 1970s.

> It often happens that the requirements of simplicity and beauty are the same, but where they clash, the latter must take precedence.
> Paul Dirac, 1939

Today, there's been a lot of progress towards unifying the weak and strong forces and electromagnetism. But the greater goal of including gravity still eludes us.

The condensed idea
Force carriers

23 The Lamb shift

What does an electron look like? An answer to this question in the late 1940s enabled physicists to fix a problem with the mathematics describing a quantum view of electromagnetism. The electron is blurred by interactions with field particles, so it appears to have a finite size.

In the 1930s, physicists knew quite a lot about electrons, Niels Bohr's simple 1913 model, which treated electrons like negatively charged planets circling a positively charged nucleus, had been upgraded to account for the shielding of the outer electrons by inner ones and the effects of angular momentum. Energy shifts due to electron 'spin' in the spectral lines of hydrogen showed that electrons act like rotating balls of charge.

The Zeeman and Stark effects – the fine splitting of the spectral lines of hydrogen due to magnetic and electric fields – revealed the magnetism associated with spinning electrons. And Pauli's exclusion principle explained why electrons, as fermions, can only have certain quantum properties, and how they fill up successive shells around atoms. Paul Dirac and others incorporated relativistic corrections.

Questions remained. In particular, it wasn't clear what an electron looked like. Schrödinger's wave equation described the probability that the electron lay in certain places, formulated as a wavefunction. But electrons were obviously localized in some sense, as their charges could be isolated and they could be fired at metal plates. In the emerging equations of quantum field theory, it was impossible to assign a charge or a mass to something infinitely small. But if a charged particle like an electron had a size, then how could it exist without its self-repulsion ripping it apart? The equations were full of infinities – mathematical singularities – that made them impossible to handle.

Quantum breakthrough

In 1947, an experiment came up with a clue that pushed quantum physics on to the next level. At Columbia University in New York, Willis Lamb and his student Robert Retherford found a new effect in the spectral lines of hydrogen. Having worked on microwave technology in the Second World War, Lamb sought to apply it to look at hydrogen at wavelengths much longer than visible light.

At the microwave frequencies they used, the hydrogen emission spectrum probed two particular orbitals: one spherical (called an S-state) and one more elongated (a P-state). Both had energies just above the lowest, or ground, state. Atomic theory at the time predicted that the two orbitals should have the same energy, but because they had different shapes, they should respond differently to a magnetic field. An energy difference should emerge, which could be detected as a new sort of splitting in the spectral lines of hydrogen. While it could affect any orbitals with differing shapes, the effect was much easier to see using microwaves than in the optical or ultraviolet parts of the spectrum.

That energy difference is exactly what Lamb and Retherford found. The pair crashed a beam of electrons into a beam of hydrogen atoms, at right angles to one another. Some of the electrons in the hydrogen atoms gained energy as a result, and moved up into the S orbital.

Hans Bethe (1906–2005)

Born in Strasbourg, now in France but then part of the German Empire, Hans Bethe showed an early flair for mathematics. Also a keen writer, he had an odd habit of writing forwards and then backwards on alternate lines. Bethe chose to study physics at the University of Frankfurt because 'mathematics seemed to prove things that are obvious'. He went to Munich, and completed his PhD in electron diffraction by crystals in 1928. Moving to Cambridge, Bethe's humour came to the fore when he published (and later retracted) a hoax paper on absolute zero to tease his colleague Arthur Eddington. During the war, Bethe (who was of Jewish heritage) moved to the United States, and he remained at Cornell University for the rest of his career. He worked on nuclear research and the Manhattan Project and solved the problem of how stars shine by proposing fusion reactions. This won him the Nobel Prize. Bethe's sense of fun prevailed when he lent his name to a famous paper that is now known as the 'alpha, beta, gamma' paper, by R. Alpher, H. Bethe, and G. Gamow.

Quantum rules forbade them to lose that energy by dropping into the lowest energy state, so they remained excited. The energized atoms were then passed through a magnetic field – producing the Zeeman effect – to finally land on a metal plate, where the electrons were released, generating a little current.

Microwaves (at frequencies close to that of a microwave oven) were also fired at the atoms in the magnetized region. By varying the strength of the magnetic field, Lamb was able to make electrons jump into the asymmetric P-state. These could fall down to the ground state, as quantum rules allowed this transition, before they hit the detecting plate, thus producing no current.

By noting when this happened for a number of frequencies, Lamb made a plot from which he could infer the energy shift between the S- and P-states in the absence of a magnetic field – known as the Lamb shift. It was not zero. So the theory of electrons must be incomplete.

In 1947, this finding shook up the quantum physics community. It was the hot topic of conversation at a conference that year at Shelter Island, on Long Island, New York. What did this energy shift mean in terms of the shape of the electron? And how could the equations be fixed to reflect it?

Many physicists assumed that the shift was a result of the 'self-energy' problem – due to the fact that the electron's own charge produces an electric field, in which it sits. But the equations couldn't handle this – they predicted that a free electron had an infinite mass, and that the spectral lines that resulted were all shifted by infinity in frequency. These factors of infinity haunted quantum physics.

Some means was needed to explain why the electron's mass was fixed and not infinite. Hans Bethe, while travelling home from the conference, hit upon a way to work around the problem. Realizing that a pure fix was beyond current understanding, he reworked the equations so that the electron's properties were no longer expressed in the usual terms of charge and mass but in rescaled versions of them. By choosing appropriate parameters, he was able to cancel out the infinities – an approach called renormalization.

We need science education to produce scientists, but we need it equally to create literacy in the public.
Hans Bethe in *Popular Mechanics* (1961)

The infinity problem arises out of the quantum graininess of the electromagnetic field. The electron is being jostled by the field's constituent particles, rather as Brownian motion scatters molecules in air. So the electron gets blurred out into a sphere. The blurry electron feels less attraction to the nucleus at close distances than it would if it were point-like, so the S-orbital in Lamb's experiment is raised in energy a little. The P-orbital is larger and affected less, as the electron isn't as close to the nucleus, so its energy is lower than the S-orbital's.

> What we observe is not nature itself but nature exposed to our method of questioning.
> Werner Heisenberg,
> *Physics and Philosophy* (1958)

Bethe's explanation fitted the experimental results of Lamb really well and came at just the right time to move the field of quantum physics forward. His technique of renormalization is still used, although some physicists worry it is rather ad hoc.

The condensed idea
Electrons' Brownian motion

24 Quantum electrodynamics

Quantum electrodynamics (known as QED) is the 'jewel of physics', according to one of its founding fathers Richard Feynman. Perhaps the most accurate theory known, it has brought physicists to an exceptional understanding of the behaviour of electrons, photons and electromagnetic processes.

QED is the quantum field theory of the electromagnetic force. It explains how light and matter interact and includes the effects of special relativity. Today's version describes how charged particles interact by exchanging photons and explains all the fine structure in the spectral lines of hydrogen, including those due to electron spin, the Zeeman effect and the Lamb shift.

The first shoots of QED came from Paul Dirac's attempts in the late 1920s to explain how an electron emits or absorbs a photon as it loses or gains energy in a hydrogen atom, so producing the observed set of spectral lines. Dirac applied Max Planck's idea of energy quanta to the electromagnetic field. Dirac thought of the quanta as tiny oscillators (vibrating strings or standing waves). He introduced the idea of particle interactions, during which particles could be spontaneously created and destroyed.

Breakthrough

For a decade, physicists tweaked this theory, but thought they had done all they could. Then came the realization that it only really worked for the simple case of the hydrogen atom. In situations beyond – for electrons with greater energies, or in larger atoms – the calculations broke down, requiring the electron's mass to grow infinitely large. Doubts were cast on the value of the entire theory: was quantum mechanics incompatible with special relativity? Subsequent findings in the 1940s, like the Lamb shift and electron spin, only added pressure.

Hans Bethe's 1947 reworking of the equations – using renormalization to cancel out the infinities – and his explanation of the Lamb shift saved the day. He still didn't have a fully relativistic theory though. Over the next few years, Bethe's ideas were developed by physicists including Sin-Itiro Tomonaga, Julian Schwinger and

Feynman. By further massaging the equations they managed to banish the infinities completely, winning the trio the Nobel Prize in 1965. Renormalization is still in the quantum physics canon today, but its physical meaning isn't understood. Feynman never liked it: he called it 'hocus-pocus'.

Feynman diagrams

The equations of QED are complicated. So Feynman, a zany, larger-than-life character with great imagination and a flair for teaching, came up with his own shorthand. Rather than using algebra, he simply drew arrows to represent particle interactions, following a few rules.

A plain arrow represents a particle moving from one point to another; a wavy line is used for a photon, and other force-carriers have squiggly variants. Every particle interaction can be shown as

Richard Feynman (1910—88)

Born and raised in New York, Richard Feynman was apparently a late talker. Having not spoken a word until he was three, he made up for it in his later life as a renowned lecturer and brilliant physicist. Feynman studied physics at Columbia University and then Princeton, and was invited to work as a junior scientist on the Manhattan Project at Los Alamos. Feynman was a prankster and liked to play jokes on his colleagues in the New Mexico desert. He broke into people's filing cabinets by guessing obvious lock combinations, such as the natural log $e = 2.71828 \ldots$, and left notes behind. He took up drumming and dancing in the desert – rumours developed of an 'Injun Joe'. After the war, Feynman eventually moved to Caltech, in part due to the warm weather. Known as 'the great explainer', Feynman was a supreme teacher and wrote a famous set of books encapsulating his lecture series. As well as QED, for which he won the Nobel Prize, Feynman worked on theories of the weak nuclear force and superfluids. In a famous talk, 'There's plenty of room at the bottom', he set the foundations of nanotechnology. Described by his colleague Freeman Dyson as 'half-genius, half-buffoon', Feynman later became 'all-genius, all-buffoon'.

three arrows meeting at a point, or vertex. Sequences of interactions could be built up by adding further units.

For example, an electron and positron colliding and annihilating to produce energy in the form of a photon would be drawn as two arrows meeting at a point, from which a wavy photon line emerges. Time runs left to right on the page. Because antiparticles are equivalent to real particles moving backwards in time, the positron arrow would be drawn pointing backwards, from right to left.

Two or more triple vertices may be combined to show a series of events. The photon created by that electron–positron interaction could then spontaneously disintegrate to form another particle–antiparticle pair, drawn as two further emerging arrows.

All sorts of interactions can be described using the diagrams, which work for any of the fundamental forces described by field theories – notably electromagnetism and weak and strong nuclear forces. There are a few rules that must be followed, such as the conservation of energy. And particles like quarks that cannot exist on their own must be balanced so that the incoming and outgoing particles are real entities, like protons and neutrons.

Probability variations

These diagrams are not just visual sketches: they have deeper mathematical meaning – they can also tell us how probable the interactions are. To find this out you need to know how many ways there are of getting there. For any starting and end point, the number of alternative interaction paths can be quickly ascertained by plotting all the variants. Count them up and you have the answer to what is most likely to happen.

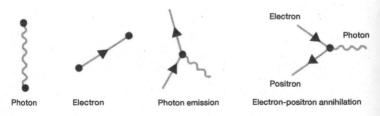

Photon   Electron   Photon emission   Electron-positron annihilation

Feynman diagrams

This influenced Feynman's thinking behind QED. He thought back to an old optics theory called Fermat's principle for the propagation of light. In working out the path of a beam of light through a lens or prism, where it may be bent, that theory states that while light rays may follow all possible paths it is the quickest path that is most probable, and where most of the light travels in phase. By counting his diagrams, Feynman also looked for the most likely outcome in a quantum interaction.

QED led the way for further developments in quantum field theory. Physicists extended this picture to cover the colour force field of quarks, a theory called quantum chromodynamics, or QCD. And QED was merged with the weak nuclear force into a combined 'electroweak' theory.

> Quantum electrodynamics (QED) has achieved a status of peaceful coexistence with its divergences . . .
> Sidney Drell, 1958

The condensed idea
Fully fledged electromagnetism

25 **Beta decay**

Unstable nuclei sometimes break down, releasing energy as particles. Beta decay occurs when a neutron becomes a proton and emits an electron plus an antineutrino. Enrico Fermi's 1934 theory of beta decay still prevails, and set the scene for studies of the weak nuclear force, which holds protons and neutrons together within the nucleus.

Radioactivity emanates from the nucleus of an atom, through the weak nuclear force. It comes in three types – alpha, beta and gamma. Alpha particles are bare helium nuclei, comprising two protons and two neutrons, and are emitted when the unstable nucleus of a heavy radioactive element such as radium or uranium breaks down. Beta particles are electrons released from the nucleus when a neutron decays into a proton. Gamma rays are energy released as photons.

Because alpha particles are relatively heavy they don't travel far and can easily be stopped by a piece of paper or our skin. Beta particles are light and travel further – it takes lead or a thick piece of metal to stop them. Gamma rays are more pervasive still.

In experiments similar to those used previously to identify the electron, in 1900 Henri Becquerel measured the ratio to mass of a beta particle's charge and found that it matched that of an electron. In 1901, Ernest Rutherford and Frederick Soddy noticed that beta radiation changed the nature of the chemical element from which it came, moving it one place to the right in the periodic table. Caesium, for example, turns into barium. So, they concluded, beta particles must be electrons from the nucleus.

In 1911, the German scientists Lise Meitner and Otto Hahn found a puzzling result. Whereas alpha particles were given off only at particular energies, beta particles could take on any amount of energy, up to some maximum. It seemed that some energy, which should have been conserved, was disappearing somewhere.

Quest for a missing particle

Nor was momentum being conserved – the direction and velocity of the nuclear recoil and the emitted beta particle did not counterbalance one another. The best explanation was that some other particle was being given off, dumping the spare energy and momentum. But nothing obvious was detected.

In 1930, in a famous letter that began 'Dear radioactive ladies and gentlemen', Wolfgang Pauli proposed the existence of an extremely light neutral particle, a companion to the proton, in the nucleus. He dubbed it the neutron. It was later renamed the neutrino (meaning 'little neutral one') by Enrico Fermi, to avoid confusion with the heavier neutron that was discovered by James Chadwick in 1932.

This light particle, Pauli thought, could explain the discrepancies, and yet, having no charge and little mass, it would be easy for it to have escaped detection. In 1934, Fermi published a full theory of beta decay, including the properties of the invisible neutrino. It was a tour de force, but Fermi was devastated when it was rejected by the scientific journal *Nature* on the basis that it was too speculative. For a while he switched his research to other topics.

> Beta decay was . . . like a dear old friend. There would always be a special place in my heart reserved especially for it.
> Chieng-Shiung Wu

Neutrinos

In fact neutrinos barely interact with matter at all, and it took until 1956 to spot them. Clyde Cowan and his collaborators turned protons and antineutrinos from beta decays into positrons and neutrons.

Enrico Fermi (1901–54)

As a child in Rome, Enrico Fermi took an interest in science, dismantling engines and playing with gyroscopes. When his father died while he was in his teens, he immersed himself in study. While studying physics at university in Pisa, Fermi became so good at quantum physics that he was asked to organize seminars, and in 1921 he published his first paper on electrodynamics and relativity. He received his doctorate aged just 21, and a few years later became a professor in Rome. Fermi's theory of beta decay was published in 1934, but, disappointed at the lack of interest in it, he turned to experimental physics, performing early work on neutron bombardment and fission. After he won the Nobel Prize in 1938 for his nuclear work, he moved to the United States to avoid the fascist regime of Benito Mussolini. Fermi's group generated the first nuclear chain reaction in Chicago in 1942, and he joined the Manhattan Project. Known for his clear and simple thinking and abilities in both experimental and theoretical physics, Fermi was one of the greatest physicists of the 20th century. The writer C.P. Snow remarked on his talents: 'Anything about Fermi is likely to sound like hyperbole.'

(For reasons of quantum symmetry the particle emitted during beta decay is actually an antineutrino.)

Neutrinos are still hard to detect. Because they carry no charge they do not ionize anything. Because they are so light they leave little trace when they hit a target. In fact most of them travel straight through the Earth.

Physicists can detect the occasional one that's slowed down by looking for flashes of light as they traverse large bodies of water – from giant swimming pools to the Mediterranean and the Antarctic ice sheet. The incoming neutrino may knock a water molecule and prompt an electron to pop out, which produces a streak of blue light (known as Cerenkov radiation).

In 1962 Leon Lederman, Melvin Schwartz and Jack Steinberger showed that there are other types (called flavours) of neutrino, when they detected interactions of the muon neutrino, a heavier member of the family than the electron neutrino. The third type, the tau neutrino, was predicted to exist in 1975 but was not seen until 2000 at Fermilab.

Neutrinos are produced by some fusion reactions that power our Sun and other stars. In the late 1960s, physicists trying to detect neutrinos from the Sun realized that they saw too few: only 30–50% of the expected number was getting through.

The solar neutrino problem was not solved until 1998, when experiments such as Super-Kamiokande in Japan and the Sudbury Neutrino Observatory in Canada showed how neutrinos change – or oscillate – between the three flavours. The relative numbers of electron, muon and tau types were being wrongly assessed earlier and the various detectors were missing some types. The neutrino oscillations indicate that neutrinos have a small mass.

> Once basic knowledge is acquired, any attempt at preventing its fruition would be as futile as hoping to stop Earth from revolving around the Sun.
> Enrico Fermi. 'Atomic Energy for power', *Collected Papers (Note e Memorie)*

So, by solving the problem of beta decay, Pauli and Fermi opened up a new world of electron-like substitutes – called leptons – as well as predicting the existence of the neutrino, a particle whose properties are still puzzling today. It set the scene for investigations of the nuclear forces.

The condensed idea
Mysterious missing particle

26 Weak interaction

The weakest of the fundamental forces, the weak force governs the decay of neutrons into protons and affects all fermions. One of its odd properties is that it is not mirror-symmetric – the universe is left-handed.

The weak nuclear force causes radioactive decay. Most particles, even the neutron, eventually break down into their fundamental constituents. While stable and long-lived within an atomic nucleus, free-flying neutrons are unstable, transforming within only around 15 minutes into a proton, electron and antineutrino.

Neutron decay underlies beta radiation. It makes radiocarbon dating possible – the carbon-14 isotope decays through the weak interaction to become nitrogen-14, with a half-life of some 5,700 years. And in the opposite sense, the weak interaction makes nuclear fusion possible, building deuterium and then helium from hydrogen within the Sun and other stars. So heavy elements are produced using the weak interaction.

The weak force is so called because its field strength is millions of times less than the strong nuclear force, which binds protons and neutrons within the nucleus, and it is thousands of times weaker than the electromagnetic force. Whereas the electromagnetic force can traverse a large distance, the weak force acts over a tiny range – about 0.1% of the diameter of a proton.

Beta decay

In the 1930s, Enrico Fermi developed his theory of beta decay and began to disentangle the properties of the weak force. Fermi saw parallels between the weak force and electromagnetism. Just as charged particles interact by exchanging a photon, the weak force must be transmitted by similar particles.

Physicists turned to the basics. What is a neutron? Werner Heisenberg imagined that the neutron was a combination of a proton with an electron stuck on it, like a molecule. He thought that larger nuclei and combinations were held together by a sort of chemical bond, with protons and neutrons bound by the exchange of electrons. In a series of papers in 1932 he tried to explain the stability of the helium nucleus (two protons and two neutrons bound together) and other isotopes. But his theory didn't pan out – within a few years experiments showed that it could not explain how two protons could hold together or interact.

Physicists looked to symmetry. In electromagnetism charge is always conserved. When particles decay or combine, charges may add or cancel but they aren't created or destroyed. Another conserved property in quantum mechanics is 'parity': the symmetry under reflection of the wavefunction. A particle has even parity if it doesn't change when it is reflected from side to side or up and down; otherwise it has odd parity.

But things were not so clear with the weak force. In fact in 1956 Chen Ning Yang and Tsung-Dao Lee proposed the radical possibility that parity might not be conserved in weak interactions. In 1957 Chieng-Shiung Wu, Eric Ambler and their colleagues at the US Bureau of Standards in Washington, DC devised an experiment to measure the parity of electrons given off in beta decay. Using cold cobalt-60 atoms, they passed the beta particles that emerged through a magnetic field. If parity was even and the electrons emerged with random orientations, then a symmetric pattern would result. If they had a preferred direction, an asymmetric pattern should appear.

Parity violation

Physicists keenly awaited the results. Wolfgang Pauli was so convinced that symmetry would be conserved that he said he was willing to bet a large sum on the outcome, declaring: 'I do not believe that the Lord is a weak left-hander.' Within a fortnight Pauli ate his words – parity was not conserved.

Later that year Maurice Goldhaber and his team at Brookhaven National Laboratory established that the neutrino and antineutrino have opposite parity – the neutrino is 'left-handed', the antineutrino 'right-handed'. The weak force, it was postulated, acted only on left-handed particles (and right-handed antiparticles). Today we know of many more particles and the picture has grown more complicated; nevertheless the point stands firm that parity is broken in weak interactions.

A flurry of theorists turned to the problem. In November 1957, Julian Schwinger proposed that three bosons were involved in transmitting the weak force. To pass on charge they must include a pair with opposite charge: he named them W^+ and W^- The third particle would have to be neutral. He assumed it was the photon. In beta decay, he reasoned, the neutron would decay into a proton and W^-, which would in turn decay to become an electron and antineutrino.

A decade earlier, Julian Schwinger had wondered if the restricted range of the weak force meant that its force-carrier was massive. The photon has no mass so can travel very far. But the weak force equivalent might be so heavy that it can't make it beyond the nucleus. The W bosons must be heavy and short-lived, their almost immediate decay explaining why we hadn't yet seen them.

Schwinger set his graduate student Sheldon Glashow to work. Glashow took his time but went one better. He realized that the fact that the W particles carried charge meant that the weak force and electromagnetism were linked. In the next few years he prepared a new theory linking the two, but it required that the third neutral particle was also massive – it was named the Z^0. So the weak force was carried by three heavy bosons: W^+, W^- and Z^0.

> There is only one thing worse than coming home from the lab to a sink full of dirty dishes, and that is not going to the lab at all!
>
> Chieng-Shiung Wu, quoted in 2001

By 1960, Glashow's theory was advanced but it was languishing. Just as quantum electrodynamics had been, it was riddled with infinities, and no one could figure out how to cancel them out. Another problem he wrestled with was explaining why the weak force carriers had large masses, whereas the photon had none.

Electroweak theory

A solution for 'electroweak' theory, combining the weak force and electromagnetism, awaited a better understanding of protons and neutrons, and the fact that they are made up of smaller particles called quarks. The weak force changes quarks from one type – or flavour – into another. Changing a neutron into a proton involves swapping the flavour of one quark.

The mass problem was solved theoretically in 1964 when a new sort of particle – the Higgs boson – was proposed. Its discovery was reported in 2012. It attracts and effectively puts a drag on the W and Z bosons, giving them their inertia. Because the W and Z bosons are so massive, weak decays are relatively slow. Hence it takes minutes for a neutron to decompose, whereas photons are given off in a fraction of a second.

Since the beginning of physics, symmetry considerations have provided us with an extremely powerful and useful tool in our effort to understand nature.
Tsung-Dao Lee, 1981

Around 1968, Glashow, Abdus Salam and Steven Weinberg presented a unified theory of the electroweak force, for which they won the Nobel Prize. Martinus Veltman and Gerard 't Hooft managed to renormalize the theory, and lose the infinities. Evidence for the W and Z particles followed in accelerator experiments in the 1970s, and they were directly detected at CERN in 1983.

While the laws of nature were long thought to be symmetric under mirror reflections, the weak force is not. It has a 'handedness'.

The condensed idea
Left-handed force

27 Quarks

n trying to explain the variety of elementary particles, Murray Gell-Mann discovered patterns that could be explained if the particles were each made from a trio of more basic components. Inspired by a literary passage, he named them quarks. Within a decade quarks were found to exist.

By the 1960s physicists had discovered around 30 elementary particles. As well as electrons, protons, neutrons and photons, there were dozens of more exotic ones with names like pions, muons, kaons and sigma particles – and all their antiparticles too.

Enrico Fermi apparently once remarked: 'If I could remember the names of these particles I would have been a botanist.' The search began for a sort of periodic table of the particles, through which to link them.

Particles fall into two basic types. Matter is made of fermions, which divide into two further sorts: leptons, including electrons, muons and neutrinos; and baryons, including protons and neutrons. Forces are carried by bosons, including the photon and various 'mesons', such as the pions and kaons responsible for the strong force.

The Eightfold Way

While visiting the Collège de France in Paris – and allegedly drinking rather a lot of good red wine – Murray Gell-Mann tried to piece the quantum properties of all these particles together. It was like solving a giant sudoku puzzle. When he grouped them by quantum traits, such as their charge and spin, a pattern began to emerge. He found a similar arrangement could explain two sets of eight particles (baryons with spin ½ and mesons with spin 0). In 1961 he published his vision as the 'Eightfold Way', named after the Buddha's eight steps to Nirvana.

One of the mesons was missing though – only seven were then known. So he stuck his neck out and predicted the existence of an eighth meson. It was found just a few months later by Luis Alvarez and his team at the University of California at Berkeley. When a new trio of spin – ³⁄₂ bosons was discovered soon after, Gell-Mann found he could fit them into another set that included ten entities. The pattern was taking shape.

Each arrangement made sense mathematically if there were three fundamental particles at the root of all of these patterns. If protons and neutrons were each made of three smaller particles, then you could rearrange the components differently to build up these particle family trees.

The basic units would have to have an unusual charge, of plus or minus $\frac{1}{3}$ or $\frac{2}{3}$ that of the electron, so that their combination gave the proton's single positive charge, or zero for a neutron. Such fractional charges seemed ridiculous – nothing like them had ever been seen – so Gell-Mann gave the imaginary particles a nonsense name, 'quorks' or 'kworks'.

Quarks and their flavours

While reading James Joyce's *Finnegans Wake*, Gell-Mann found a better name in one passage: 'Three quarks for Muster Mark!' Joyce's word referred to the squawk of a seagull, but Gell-Mann enjoyed the similarity to his own made-up word, and its link to the number 3. In 1964 he published his quark theory, proposing that the neutron is a mixture of an 'up' and two 'down' quarks, and a proton is two ups and a down. So beta radiation, he declared, involved the conversion of a down quark in a neutron into an up quark in a proton, while emitting a W⁻ particle.

Gell-Mann's magical Eightfold Way seemed to work, but even the physicist himself didn't know why. He accepted it was merely a mathematical device. Others treated his quark theory with derision at first. There was little evidence for quarks' physical existence, until experiments at the Stanford Linear Accelerator Center in 1968 revealed that the proton was indeed made up of smaller components.

Today, as more and more particles have been discovered, Gell-Mann's picture has been vindicated. We know that there are six types, or flavours, of quarks: up, down, strange, charm, bottom and top. These come in pairs; the up and down quarks are the lightest and most common. Evidence for the heavier quarks is visible only in high-energy collisions – the top quark was not spotted at Fermilab until 1995.

The weird names of quarks and their characteristics have arisen in an ad hoc way. The up and down quarks are the simplest, named after the direction of their isospin (a quantum property in the strong and weak nuclear forces analogous to charge in electromagnetism).

Three quarks for Muster Mark! Sure he has not got much of a bark. And sure any he has it's all beside the mark.

James Joyce, *Finnegans Wake*

Strange quarks are so called because they have turned out to be components of the 'strange' long-lived particles discovered decades earlier in cosmic rays. The 'charm' quark was named for the pleasure it brought its discoverer. Bottom and top were chosen as complementary to up and down. Some physicists refer more romantically to top and bottom quarks as 'truth' and 'beauty'.

Quarks can change their flavour through the weak interaction, and respond to all four fundamental forces. For every quark there's an antiquark. Particles made of quarks are called hadrons (from the Greek *hadros*, 'large'). Quarks cannot exist on their own – they only come in threes and are confined within hadrons.

Quark 'colours'

Quarks have their own set of properties, including electric charge, mass, spin and a further quantum trait known as 'colour' charge, linked to the strong nuclear force. Quark colours are labelled red, green and blue. Antiquarks have anticolours, such as anti-red. Just as, in optics, the three primary colours combine to give white light, baryons must be made up of a combination that mixes to white.

The attraction and repulsion of quarks of various colours is governed by the strong force and is mediated by particles called 'gluons'. The theory that describes strong interactions is called quantum chromodynamics (QCD).

How can it be that writing down a few simple and elegant formulae, like short poems governed by strict rules such as those of the sonnet or the waka, can predict universal regularities of Nature?

Murray Gell-Mann, Nobel Banquet Speech
(10 December 1969)

The condensed idea
the power of three

28 Deep inelastic scattering

A series of experiments in California in the late 1960s confirmed the quark model of the proton and other hadrons. By firing electrons with very high energies at protons, physicists showed that they rebounded strongly when they hit three points within the nucleon and that the quarks had fractional charge.

In 1968, physicists at Stanford University puzzled over results from their new particle accelerator. The Stanford Linear Accelerator Center (SLAC), just south of San Francisco, wasn't the highest-energy particle smasher in the US – that was at Brookhaven on the east coast. But SLAC was built to perform a bold feat – to tear apart the proton.

The bigger accelerators of the day, like Brookhaven, mostly collided beams of weighty protons, looking for new types of particles among the shards of the smash-ups. Richard Feynman famously referred to this as like smashing apart a Swiss watch to find out how it worked. The SLAC team instead fired fast beams of electrons at protons.

While electrons are much lighter than protons, and so should have less impact, the American theorist James Bjorken realized that they can do more precise damage. Very high-energy electrons would have very compact wavefunctions. The electrons would target their blow on a region tiny enough to pierce the proton. In essence the SLAC physicists were going one step further than Ernest Rutherford, who 50 years earlier had discovered the atomic nucleus by firing alpha particles at gold foil.

In the 1960s, physicists didn't know what protons were made of. Murray Gell-Mann had proposed they might be built of three quarks, but this idea was purely conceptual: it cut no ice in experimental circles. Just as Rutherford first imagined a 'plum-pudding' atom, so the proton might be a ball of some smeared-out substance. Or, like Niels Bohr's atom, it might be mostly empty space inhabited by tiny constituents.

Two kinds of collision
In the SLAC accelerator an electron could collide with a proton in two ways. In the simplest case it could rebound off the nucleus, both particles remaining intact and responding according to the

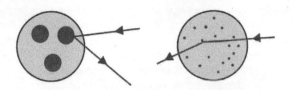

Quarks inside a proton scatter incoming electrons, which would otherwise pass through.

conservation of momentum. Because kinetic energy is not lost, this is described as an elastic collision. Alternatively, the electrons might undergo inelastic collisions, where some of the kinetic energy gets turned into new particles.

Inelastic collisions might be modest, where the proton essentially remains in place, absorbing some energy from the electron plus also creating some other particles as splinters. On the other hand the electron might pierce the proton and blow it apart – the innards blasted out in a much larger shower of particle shrapnel. This more destructive process is known as 'deep inelastic scattering'. Bjorken realized it could reveal how the proton is built.

If the proton is a smooth mass, then after the collision the electrons should be deflected only a little from their incoming path. If the proton is made of tiny hard cores, then the lightweight electrons could bounce off them at quite large angles, just as Rutherford witnessed alpha particles ricocheting off hard gold nuclei.

Bjorken's team quickly saw that many of the electrons were widely deflected. And they spotted peaks in the relative energy of the scattered electrons, suggesting underlying structure in the proton. Protons must be made up of tiny grains.

Physicists also collide

The interpretation of the grains as quarks didn't follow immediately. There were other possibilities. Richard Feynman, fresh from collecting his Nobel Prize for his work on quantum electrodynamics, promoted a different model. He too wondered if protons and other hadrons were made of smaller components, but his version he called 'partons' (parts of hadrons).

Feynman's model was still at an early stage. He didn't know what partons were, but imagined how they would clash together during collisions if the proton and electron were flattened as they experienced relativistic effects. Feynman was convinced that the SLAC results supported his parton model and, given his popularity and recent award, for a time many Californian physicists were happy to believe him.

But further experiments started to back the quark model. Neutrons became the next targets, and produced a subtly different pattern in the scattered electrons, implying their make-up was slightly different. It took several years and lots of wrangling to agree what the defining tests were and how to interpret the data, but in the end the quark model passed.

I believe there are 15,747,724,136,275,002,577,605,653,961,181,555, 468,044,717,914,527,116,709,366,231,025,076,185,631,031,296 protons in the universe, and the same number of electrons.
Sir Arthur Stanley Eddington, 1938

Protons and neutrons and other baryons have three scattering centres within them, corresponding to three up and down quarks. Mesons have two scattering points, corresponding to a quark and an antiquark. The grains are extremely compact – essentially point-like, like the electron. And they have charges of multiples of ⅓, consistent with the quark concept.

In 1970 Sheldon Glashow added a further affirmation when he deduced the charm quark's existence from the decay of heavier 'strange' particles such as the kaon. By 1973, most particle physicists accepted the quark theory.

A couple of puzzles remained: during the collisions the quarks seemed to behave like independent particles within the nucleus, but they could not be set free. Why? What was the quantum glue that held them there? And if quarks were fermions, how come two similar ones could exist side by side within a proton or neutron? Pauli's exclusion principle should rule that out.

> One could say that physicists just love to perform or interpret scattering experiments.
> Clifford G. Shull, 1994

The answers would come from the next development in quantum field theory – quantum chromodynamics (QCD), or the study of the varied properties of quarks and the strong force that governs them.

The condensed idea
The core of things

29 Quantum chromodynamics

With the confirmation of quark theory, the search began for a fuller explanation of the strong interaction that dictates the behaviour of protons and neutrons in the nucleus. Quantum chromodynamics (QCD) describes how quarks experience a 'colour' force, which is mediated by gluons.

In the 1970s, physicists began to accept that protons and neutrons were made up of a trio of smaller components called quarks. Originally predicted by Murray Gell-Mann to explain patterns he perceived in the characteristics of elementary particles, quarks had some weird properties. Experiments at Stanford Linear Accelerator Center revealed the graininess of protons in 1968, and later the same for neutrons, by firing fast electrons at them. Quarks have charges that are plus or minus ⅓ or ⅔ the basic unit, so that three of them add together to give the +1 charge of the proton or 0 for the neutron.

Within the SLAC experiments, the quarks behaved as if they were disconnected. But they could not be extricated from the nucleus – they had to be confined within it. No particles with fractional charges were ever seen outside. It was as if they were rattling around inside the proton, like beans in a shaker. What was holding them in?

A second problem was that quarks are fermions (with spin ½). Pauli's exclusion principle says that no two fermions can have the same properties. Yet protons and neutrons host two up or two down quarks. How was that possible?

Colour charge
In 1970 Gell-Mann thought about these problems while he was spending the summer in the mountains of Aspen, Colorado, at a physics retreat. He realized that the exclusion principle problem could be solved if he introduced another quantum number (like charge, spin and so on) for quarks. That property he called 'colour'. Two up quarks, for example, could sit within a proton if they had different colours.

Quarks, he postulated, have three colours: red, green and blue. So the two similar up and down quarks in protons and neutrons have different colours, and Pauli's principle is preserved. A proton, for instance, could contain a red and a blue up quark and a green down quark. Because colour applies only to quarks, not to real particles like protons, the overall colour of a real particle must be white – by analogy with the colours of light. So a triple quark combination must include red, green and blue. Antiparticles have equivalent 'anti-colours'.

In 1972, Gell-Mann and Harald Fritzsch brought the three quark colours into the model of the Eightfold Way. As well as the three flavours and colours, the picture demanded eight new force carriers, to

To me, the unity of knowledge is a living ideal and goal.
Frank Wilczek, 2004

transmit the colour force. These were called gluons. Gell-Mann presented his model casually at a conference in Rochester, New York. But he still had his doubts that quarks, let alone colour and gluons, were real.

Asymptotic freedom

The harder problem to solve was that of the confinement of quarks within the nucleus. The SLAC experiments showed that the closer together they are, the more independently they behave. Conversely, when they are far apart they tug upon one another more.

This behaviour is known as 'asymptotic freedom', as at zero separation they should theoretically be completely free, and not interacting with one another. Quite opposite to forces like electromagnetism and gravity, whose strength falls off with distance, this aspect of the strong force was counter-intuitive at best.

In 1973, David Gross and Frank Wilczek and independently David Politzer managed to extend the framework of quantum theory to explain asymptotic freedom. Gell-Mann and his colleagues developed their work further, and made predictions about small discrepancies in the scattering experiments that were seen at SLAC. The entirely conceptual quark theory was – remarkably – holding true.

The new theory needed a name, and the following summer Gell-Mann came up with one: quantum chromodynamics, or QCD. It had 'many virtues and no known vices', Gell-Mann said.

No solo quarks

However, the theory wasn't quite complete. It didn't explain why quarks were never seen in isolation, why they were locked inside hadron nuclei. Physicists duly came up with an explanation. As quarks are dragged away from the core of the proton, the colour force increases and the gluons holding them back are stretched out into strings, a bit like strands of chewing gum.

If the quark continues to try to escape, the string eventually snaps, and the gluon energy is converted to quark–antiquark pairs. The escaping quark may be mopped up by the antiquark, becoming absorbed into a real particle like a meson. The other free quark stays within the nucleus. Individual quarks can never escape the colour force. Unlike photons, which carry no electric charge, gluons carry 'colour charge' and can interact with one another. In colour interactions a whole series of particles may be created out of quark–antiquark pairs, and they tend to fly off in roughly the same direction. Observations of these 'gluon jets' confirmed evidence for gluons in 1979.

In subsequent years, further quarks were found: the charm quark in 1974, the bottom quark in 1977, and finally the top quark in 1995.

QCD joined the table of other accurate quantum field theories. What remains to be discovered is a way of unifying the main three forces – electromagnetism, and the weak and strong force – to explain the standard model of particle physics.

We called the new [fourth] quark the 'charmed' quark because we were pleased, and fascinated by the symmetry it brought to the subnuclear world.
Sheldon Lee Glashow, 1977

The condensed idea
Three colours red, green and blue

30 The Standard Model

The piecing together of a complex family tree for more than 60 fundamental particles and 20 quantum parameters is a huge achievement. Patterns hint at underlying laws of nature. Nevertheless, there may be more to add to the Standard Model of particle physics.

By the mid-1980s, physicists were putting the finishing touches to their full account of the plethora of elementary particles discovered over the past century. Whereas in the 1950s and 60s, theorists had been taken aback by what was turning up in experiments, by the 1970s accelerators were dotting the 'i's and crossing the 't's of what was becoming the Standard Model of particle physics.

From Niels Bohr's opening salvo on atomic structure, electrons had turned out to be weird probabilistic creatures, answering only to quantum mechanics and described in terms of wavefunctions. The nucleus was even stranger. A ladder of entities, from quarks held in place by gluons to massive W and Z bosons and fleeting neutrinos, combined to produce once familiar behaviour such as radioactivity.

As more and more particles emerged – first from studies of cosmic rays and then in ground-based accelerators and colliders – Murray Gell-Mann's mathematical intuition was one step ahead. His 1961 Eightfold Way expressed the underlying symmetries of families of particles, governed by their quantum numbers. The theory of quarks and quantum chromodynamics followed.

By the 1990s, all that remained to be fitted into the basic gaps in the Standard Model framework were the top quark (discovered in 1995) and the tau neutrino (discovered in 2000). The Higgs boson was more icing on the cake in 2012.

Three generations

The Standard Model describes the interactions of three generations of matter particles, through three fundamental forces, each mediated by its own force carriers. Particles come in three basic types: hadrons, such as protons and neutrons that are built from quarks; leptons, which include electrons; and bosons, like photons, associated with the transmission of forces. Each hadron and lepton has a corresponding

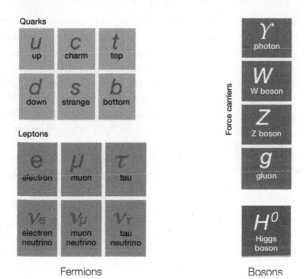

Quarks

u up	c charm	t top
d down	s strange	b bottom

Leptons

e electron	μ muon	τ tau
ν_e electron neutrino	ν_μ muon neutrino	ν_τ tau neutrino

Fermions

Force carriers

γ photon
W W boson
Z Z boson
g gluon
H^0 Higgs boson

Bosons

antiparticle as well. Quarks, too, come in threes. They have three 'colours': red, blue and green. Just as electrons and protons carry electric charge, quarks carry 'colour charge'. The colour force is transmitted by a force particle called a 'gluon'.

Rather than weakening with distance, the colour force strengthens if quarks are pulled apart, a bit like elastic. It locks them together so tightly that individual quarks can never be separated and cannot exist on their own. Any independent particle made of quarks must be colour-neutral – made up of a combination of colours that adds to white. Particles such as protons and neutrons made from three quarks are called 'baryons' ('bary' means heavy), those comprising quark–antiquark pairs are called 'mesons'.

Quarks have mass and come in six types called 'flavours'. Quarks are grouped in three generations and come in three complementary pairs. The labels are flukes of history: 'up' and 'down', 'strange' and 'charm' and 'top' and 'bottom' quarks. Up, charm and top quarks have electrical charges +⅔. and the down, strange and bottom quarks have charge – ⅓. A proton is made of two ups and a down quark; a neutron is made of two downs and an up quark.

The leptons include particles like electrons and neutrinos, which are not subject to the strong nuclear force. Like the quarks, leptons exist in six flavours and three generations with increasing masses: electrons, muons and taus and their corresponding three neutrinos (electron neutrino, muon neutrino and tau neutrino). Muons are 200 times heavier than an electron and taus 3,700 times. Neutrinos have almost no mass. Electron-like leptons have a single negative charge; neutrinos have none.

The force-carrying particles include the photon, which transmits the electromagnetic force, the W and Z particles that carry the weak nuclear force, and the gluons linked to the strong nuclear force. All these are bosons and are not subject to Fermi's exclusion principle, meaning that they can exist in any quantum state. Quarks and leptons are fermions and are restricted by Pauli's rules. Photons have no mass, gluons are light, but W and Z particles are relatively heavy. Mass is given to the W and Z bosons by another field – the Higgs field, transmitted by the Higgs boson.

Particle smashing

The discovery of the particle zoo has only been possible thanks to extreme technology. Apart from those dissected from atoms, the first exotic particles stemmed from cosmic rays, high energy particles from space that crash into Earth's upper atmosphere, sending out a shower of secondary ones that particle physicists could catch.

In the 1960s, a series of particle accelerators climbed to higher and higher energies, making it possible to create particles from scratch. By firing fast beams of protons at targets or opposing beams, new types of particles could be generated in the smash-up. High energies must be reached to create very massive particles, so the latest to be discovered tended to be the heavy generations. Lots of energy is also necessary to prise apart the strong nuclear force and temporarily release quarks.

To identify the particles, physicists pass them through a magnetic field. Positively and negatively charged particles swerve in opposite directions. Massive and light and fast and slow particles are also deflected differently, some forming tight spirals.

Outstanding questions

The Standard Model has proved remarkably resilient, and its development is certainly a huge achievement. But physicists are not crowing just yet. With some 61 particles and 20 quantum parameters, the model is cumbersome. The values of these parameters are all experimentally derived, rather than being predicted theoretically.

The relative masses of the various particles make no obvious sense. Why is the top quark so much heavier than the bottom quark, for instance? And why is the mass of the tau lepton so much greater than the electron's? The particular masses seem quite random.

The strengths of the various interactions – the relative strength of the weak and electromagnetic forces, for instance – are equally unfathomable. We can measure them, but why is it so?

And there are still gaps. The model doesn't include gravity. The 'graviton', or gravity-force-carrying particle, has been postulated, but only as an idea. Maybe one day physicists will manage to put gravity into the Standard Model – such a grand unified theory (GUT) is a huge but distant goal.

Further puzzles not yet explained by the Standard Model include some of the mysteries of the universe, among them matter–antimatter asymmetry, the nature of dark matter, and dark energy. So there's a lot left to learn.

The condensed idea
Particle family tree

31 Symmetry breaking

Physics is full of symmetries. Laws of nature remain unchanged no matter where or when we measure them. Symmetries built into most theories of physics apply to particles right across the universe. But symmetries are sometimes broken, resulting in distinctive particle masses or handedness.

We are all familiar with the concept of symmetry. The patterns on a butterfly's wings are reflections of one another; a symmetric human face is often thought beautiful. Such symmetries – or robustness to transformations – underlie much of physics. In the 17th century, Galileo Galilei and Isaac Newton assumed that the universe works the same everywhere – the same rules apply to the planets as to Earth. Laws of nature are unchanged if we move a few metres or millions of light years to the left, if we spin around or stand on our head.

Albert Einstein's special and general theories of relativity are motivated by the fact that the universe should look the same to any observer, no matter where they are or how fast they are travelling or accelerating. James Clerk Maxwell's classical equations of electromagnetism exploit symmetries between electric and magnetic fields, such that their properties can be interchanged from different vantage points. The Standard Model of particle physics also grew through considerations of symmetry. Murray Gell-Mann pieced together the jigsaw of elementary particles by finding regular patterns in the quantum numbers of particles. As a result he predicted the existence of triplets of quarks.

All three physicists – Einstein, Maxwell and Gell-Mann – developed their breakthrough theories by following their deep faith in the mathematics of symmetry. Their trust that nature followed such rules enabled them to bypass prejudices tied to existing observations and ideas to derive completely new theories, whose weird pronouncements were later found to be true.

Gauge symmetry

The quantum world is full of symmetries. Because there is a disconnect between what is observed in the real world and what is really going on below the surface, the equations of quantum mechanics and quantum field theory must be adaptable. Wave and matrix mechanics, for instance, must give the same outcome for an experiment, irrespective of how the theory is formulated. The observables – such as charge, energy or velocities – should appear the same no matter on what scale we describe the underlying field.

These laws of physics must be written in such a way that the observed quantities are unaffected by transformations in coordinates or scale (gauge). This is known as 'gauge invariance' or 'gauge symmetry', and theories that obey this are called gauge theories. As long as this symmetry hold true, physicists can rearrange the equations any way they like to explain behaviour.

Maxwell's equations are symmetric under scale transformations. So too is general relativity. But the approach was generalized most powerfully in 1954 by Chen Ning Yang and Robert Mills, who applied it to the strong nuclear force. The technique inspired Gell-Mann's search for symmetry groups of particles, and then found application in the quantum field theory of the weak force and in its unification with electromagnetism in electroweak theory.

> Symmetry, as wide or as narrow as you may define its meaning, is one idea by which man through the ages has tried to comprehend and create order, beauty and perfection.
>
> Hermann Weyl, 1980

Conservation

Symmetries are closely tied to conservation rules. If energy is conserved, then to comply with gauge invariance charge must also be conserved – as we cannot create a fixed amount of charge if we don't know what the absolute scale of a field is. When describing fields, relative effects are all that matter. Gauge symmetry explains why all particles of a given type are indistinguishable. Any two could swap position and we'd never know. Similarly, photons are inextricably linked even though they might appear distinct.

Other symmetries that are important in physics include time: the laws of nature are the same today and tomorrow and antiparticles are equivalent to real particles moving backwards in time. And parity: a measure of the symmetry of a wavefunction, such that even parity is symmetric under reflection, odd is not.

Symmetry breaking

Symmetries are sometimes broken. For example, the weak nuclear force doesn't conserve parity and prefers left-handed particles (electrons and neutrinos). Handedness (or chirality) is also a property of quarks in quantum chromodynamics (QCD), such that a left-handed particle moves and spins in the same direction. Matter and antimatter are in cosmic imbalance. And the fact that different particles have different masses requires symmetry breaking – without it, they would all be massless.

Just as water freezes quickly to become ice, symmetry breaking is rapid. At a critical point, the system clicks into a new state, which at first might seem arbitrary. An example is a pencil balanced on its point. As it stands there it is symmetric – there's an equal chance of it lying in any direction – but once it falls down it picks a compass point. The symmetry is broken.

Another example is the appearance of a magnetic field in a bar magnet. In a piece of hot iron, all the internal magnetic fields are jiggling about and oriented randomly – so the block has no overall magnetic field. But when you cool it below a threshold, known as the Curie temperature (around 700°C), the atoms undergo a 'phase transition' and most align in one direction. The cold iron gains north and south magnetic poles.

A series of similar phase transitions in the young universe explains why we have four

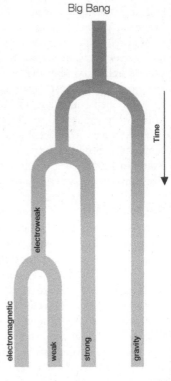

The four fundamental forces split off due to symmetry breaking in the early universe.

fundamental forces today and not just one. In the extreme heat of the very early universe, just after the Big Bang, all four forces were unified. As the universe cooled, just like the bar magnet, it underwent symmetry-breaking phase transitions.

The various forces sprang off one by one. Gravity separated first, at just 10^{-43} seconds after the Big Bang. At 10^{-36} seconds the strong interaction appeared, pulling quarks together. The weak and electromagnetic forces remained combined until around 10^{-12} seconds, when they too divided.

The energy of the universe at this electroweak phase transition is around 100 GeV. Above this energy, the W and Z bosons that carry the weak interaction and photons that transmit the electromagnetic force are indistinguishable – their equivalents are carriers of the electroweak interaction. Below this energy, though, we know that the W and Z bosons are heavy whereas the photon is massless. So they acquire their mass during the symmetry-breaking process.

Symmetry breaking explains the different masses of gauge bosons – why some are heavy, others light or having no mass. Without spontaneous symmetry breaking, all of them would be massless. The mechanism involved is known as the Higgs field, after the physicist Peter Higgs who worked on it in the 1960s.

The condensed idea
Order breakdown

32 The Higgs boson

Why are some particles more massive than others? The Higgs boson was postulated by Peter Higgs in 1964 as a way to give inertia to particles. It tugs on force carriers such as the W and Z bosons and breaks the symmetry between the weak and electromagnetic forces.

By the 1960s, it was known that the four fundamental forces were carried by very different particles. Photons mediate electromagnetic interactions, gluons link quarks by the strong nuclear force, and the W and Z bosons carry weak nuclear forces. But unlike photons, which have no mass, the W and Z bosons are heavy, weighing a hundred times more than a proton. Why do particles have a range of masses?

Physicists looked to symmetry. The Japanese-born American theorist Yoichiro Nambu and the British physicist Jeffrey Goldstone proposed that a spontaneous symmetry-breaking mechanism generates a slew of bosons during the separation of forces. Yet in their models these bosons had no mass – by implication, all the force carriers should be like the photon.

It didn't make sense. Massive force carriers are needed for short-range forces, the physicists reasoned. Massless bosons, like the photon, can travel large distances, whereas the nuclear forces are obviously localized. If the weak and strong forces have massive carriers, then their small range could be explained.

Commenting on the futility of generating force carriers from the vacuum, as Nambu and Goldstone had done, fellow physicist Steven Weinberg quoted from Shakespeare's *King Lear*: 'Nothing will come of nothing'.

Phil Anderson, a condensed-matter physicist, made a suggestion by drawing upon the behaviour of paired electrons in superconductors. Nambu and Goldstone's massless bosons should effectively cancel each other out, he reasoned, so ones with finite mass would be left.

A flurry of papers enlarging on this idea followed in 1964, written by three teams: the Belgian physicists Robert Brout and François Englert working at Cornell University, the British physicist Peter Higgs at Edinburgh University, and Gerald Guralnik, Carl Hagen and Tom Kibble at Imperial College London. The mechanism they derived is now known as the Higgs mechanism. While all three groups performed similar calculations, Higgs went on to describe the mechanism in terms of a boson – the Higgs boson.

Higgs boson

Higgs imagined the W and Z force carriers as being slowed by passage through a background force field. Now known as the Higgs field, it is mediated by Higgs bosons. By analogy, a bead dropped into a glass of water will fall more slowly than in air. It is as if the bead is more massive when in water – it takes longer for gravity to pull it through the liquid. The bead may sink even more slowly in a glass of syrup. The Higgs field acts in a similar way, like molasses.

Or imagine a celebrity walking into a cocktail party. The star will barely make it past the door before they are swamped with fans, who slow their movement across the room. The W and Z bosons are the particles with star appeal: the Higgs force acts more strongly on them than on photons, so they appear heavier.

Smoking gun

Hints of the Higgs boson were detected in 2011 but the signals were confirmed convincingly – to great fanfare – in 2012. It took decades to build a machine capable of finding the Higgs boson because of the high energies at which it ought to exist (more than 100 GeV). In 2009 the multi-billion-dollar Large Hadron Collider (LHC) at CERN, in Switzerland, began operating. CERN, the Conseil Européen pour la Recherche Nucléaire (European Organization for Nuclear Research), is a vast particle physics facility near Geneva. About 100 metres under the Swiss–French border is a 27-kilometre-long ring of tunnels, through which particle beams are accelerated by giant superconducting magnets.

Two opposing proton beams smash into each other head-on. The huge energies produced in the crash allow a range of massive particles to be released temporarily in the carnage and recorded by detectors. Because the Higgs boson is heavy it can only appear at extreme energies and, owing to Heisenberg's uncertainty principle, only for a very short time. The Higgs particle's signature must be teased out from billions of other particle signatures. It has been a difficult search.

On 4 July 2012, two experimental teams at CERN claimed to have seen a new particle with the energy expected for the Higgs boson according to the Standard Model (126 GeV). The particle's identity needs to be confirmed through further measurements, but its appearance is tantalizing. As well as being another tick for the Standard Model, it opens up a host of new questions for particle physicists to explore.

> [The Large Hadron Collider] is the Jurassic Park for particle physicists . . . Some of the particles they are making now or are about to make haven't been around for 14 billion years.
>
> Phillip Schewe, 2010

First, exactly how does the Higgs boson confer mass? From neutrinos to the top quark, there are 14 orders of magnitude of mass that the Standard Model needs to explain. And then, how does the Higgs boson get its own mass? Stay tuned.

The condensed idea
Wading through molasses

33 Supersymmetry

The Standard Model's lack of elegance has led to searches for a more basic theory of particles and physical forces. Supersymmetry supposes that every particle has a supersymmetric partner, identical apart from its quantum spin. Like antimatter, the introduction of these new particles makes quantum field equations easier to solve and more flexible.

The Standard Model has done a remarkable job of tying together the varied properties of over 60 elementary particles. Like a pricey box of chocolates, the many particles may be grouped into layers according to their style. But the Standard Model remains very complicated and physicists yearn for simplicity. There are many open questions. For example, why do so many particle properties come in sets of three? Why are there three generations of leptons – electrons, muons and taus and the corresponding neutrinos? Two generations were too many for the Nobel laureate I.I. Rabi, who exclaimed upon its discovery, 'Who ordered the muon?' Likewise, the three generations of quarks needs explaining. Why do particles have such a wide range of mass? From the electron to the top quark, fermions span six orders of magnitude in mass. The recent discovery of neutrino oscillations – showing that neutrinos have a tiny mass – pushes the mass range towards thirteen or fourteen orders of magnitude. With so many possibilities, why does any particle take the mass that it does?

The strengths of the four fundamental forces – related to the mass of their carrier particles – are also impossible to predict in the Standard Model. Why exactly is the strong force strong and the weak force weak? Then there is the Higgs boson. Its existence was deduced for the purpose of breaking symmetry in electroweak interactions. So far we only know of one Higgs boson. But could there be more particles like it? And what else might be out there? Even though there are regularities to the patterns it supports, the whole framework of the Standard Model seems ad hoc.

Beyond the Standard Model

The Standard Model's messiness suggests to some that we are not there yet – that one day we will realize that the model is a small part of a broader and more elegant theory. Physicists are again returning to basic definitions and concepts like symmetry to see what sort of qualities such an overarching theory might have.

When seeking a more fundamental basis for understanding some phenomenon physicists tend to look at ever-smaller scales. The physics of ideal gases, pressure and thermodynamics require an understanding of molecular processes, and theories of atoms require an understanding of electrons and nuclei.

Take the electron. Physicists can use the equations of electromagnetism to explain its properties some distance away from the particle, but the closer you get to it the more the electron's influence upon itself comes to dominate. As the fine structure of hydrogen spectral lines demonstrates, the charge, size and shape of the electron are important.

As the development path of quantum electrodynamics (QED) showed, it took a quantum-mechanical view of the electron as a wavefunction, including the effects of special relativity, to describe all its properties. Paul Dirac managed to write that down in 1927, but the new picture came with a major consequence the existence of antimatter. The number of particles in the universe doubled, and a host of new interactions could be considered.

The equation for electrons only makes sense if there are also positrons, with quantum properties that are the flipside of the electron. For a length of time that depends on Heisenberg's uncertainty principle, electrons and positrons may pop into existence in the vacuum of space, then annihilate. These virtual interactions solve problems such as the size of an electron, which otherwise create discrepancies in the theory.

To go beyond the Standard Model, we must consider processes on smaller scales and at higher energies than the most extreme ones currently known, namely the Higgs boson (whose energy exceeds 100 GeV). Just as with the electron, physicists must ask what a Higgs boson actually looks like and how its own shape and field affects its behaviour at close quarters.

Particle twins

Again, like the electron and positron, the solution to this physics problem requires another doubling in the number of possible particles

– such that every particle has a 'supersymmetric' partner (with the same name but an 's' prefix). The supersymmetric partner for the electron is called the selectron, quarks have squarks. The photon and W and Z boson equivalents are called the photino, wino and zino.

Supersymmetry (often shortened to SUSY) is a symmetry relationship between bosons and fermions. Every boson – or particle with integer spin – for example, has a corresponding fermion or 'superpartner' whose spin differs by ½ a unit, and vice versa. Apart from spin, all the quantum numbers and mass are the same for the superpartners.

> But although the symmetries are hidden from us, we can sense that they are latent in nature, governing everything about us. That's the most exciting idea I know: that nature is much simpler than it looks.
> Steven Weinberg, Sidelights

Although attempts were made in the 1970s, the first realistic supersymmetric version of the Standard Model was developed in 1981 by Howard Georgi and Savas Dimopoulos. For bosons, it predicts a range of superpartners with energies between 100 and 1,000 GeV, that is, just above or similar to that of the Higgs. Like the positron, the existence of these supersymmetric particles cancels out irregularities in the descriptions of particles at close range.

The lower end of this energy range is now accessible with the Large Hadron Collider at CERN. To date (as of 2022) there is no evidence for supersymmetric particles. We'll see what happens as the collider's energy is raised within a few years.

If the superpartners remain out of reach, physicists could speculate that they have even higher masses than their Standard Model partners. In that case, supersymmetry must be being broken, suggesting yet another level of particles that needs to be explored.

Ultimately, supersymmetry could help to unify the weak and strong interactions and electromagnetism and perhaps eventually also gravity. The complementary approaches of string theory and quantum gravity would have to incorporate it, especially if hints of supersymmetric particles are found.

Supersymmetry has some attractive features. The undetected superpartners are good candidates for the ghostly dark matter that haunts the universe. Dark matter makes up most of the mass of the universe but it only reveals itself through its gravitational effect, otherwise it is invisible.

The condensed idea
Spin symmetry

34 Quantum gravity

The holy grail of a theory of all four fundamental forces eludes us. But that has not stopped physicists from trying to meld together quantum theory and general relativity. Such theories of quantum gravity are a long way off but hint that space might be a fabric of tiny knitted loops.

When Albert Einstein presented his theory of general relativity in 1915 he recognized that it would have to be reconciled with the emerging quantum theory of the atom. Just as planets are held by gravity around the Sun, even electrons should experience gravitational forces as well as the electromagnetic ones that hold them in their shells. Einstein worked for much of his life to develop a full quantum theory of gravity. It eluded him – and still eludes us today.

After Einstein, Niels Bohr's protégé Leon Rosenfeld began the process in the 1930s, when quantum mechanics was put on the table. Fundamental obstacles were immediately thrown up. The first is that general relativity is not tied to an absolute backdrop, whereas quantum mechanics is.

Relativity applies most of all to massive objects, like planets and stars and galaxies and matter across the entire universe. Its equations don't distinguish space and time but treat them as four dimensions of a smooth metric called space-time. Masses move within this fabric, distorting it according to their mass. But there is no absolute grid of coordinates. As its name implies, the theory of general relativity explains relative motions, of one object relative to another in curved space-time.

By contrast, quantum mechanics does care about where and when a particle is located. Wavefunctions are dictated by, and evolve according to, the local surroundings – the wavefunction of every particle in a box or electron in an atom is different. In the quantum picture, space is not empty or uniform but a sea of virtual quantum particles, popping in and out of existence.

Just as drawing together Heisenberg's matrix mechanics and Schrödinger's wave equation was fundamentally difficult because one was discrete and the other continuous, so reconciling quantum mechanics and general relativity is like matching apples and oranges.

There are three areas where the disconnection is greatest. Both general relativity and quantum mechanics break down or become inconsistent at, or near, singularities, such as black holes. Second, because the Heisenberg uncertainty principle means a particle's location and velocity cannot be known with certainty, it is impossible to say what gravity it feels. Third, time has a different meaning in quantum mechanics and general relativity.

Quantum foam

Work on quantum theories of gravity picked up in the 1950s. The Princeton University physicist John Wheeler and his student Charles Misner applied Heisenberg's uncertainty principle to describe space-time as a quantum 'foam'. On the tiniest scales, they proposed that space-time distorts into a tangle of tunnels, strings, lumps and bumps. In 1957 Misner recognized that there are two ways forward. First, general relativity can be rewritten in a form of calculus more similar to that of quantum mechanics. Then that theory could be quantized. The alternative is to expand existing quantum field theories to include gravity, following a similar path to quantum electrodynamics and attempts to include the nuclear forces. A new force carrier for gravity would be needed – the graviton.

In 1966, Bryce DeWitt, who had studied under Julian Schwinger, took a different tack after a discussion with Wheeler. Familiar with cosmology – and the recent discovery of the cosmic microwave background radiation – DeWitt published a wavefunction for the universe. It is now known as the Wheeler–DeWitt equation. He took the equations for the expansion of the universe after the Big Bang and treated the cosmos as a sea of particles.

One odd outcome was that time was unnecessary in DeWitt's formulation. He only needed the three coordinates of space – time was just a manifestation of changing states of the universe, which we perceive as a sequence. Just as Schrödinger struggled to understand what his wave equation meant, DeWitt could not explain what his universal wavefunction was describing in real life. Whereas the Copenhagen interpretation tied the quantum to the classical world, when it came to the whole universe there was nothing to compare it with. There could be no external 'observer' whose attention collapsed the cosmic wavefunction.

Other physicists worked on the problem, including Stephen Hawking, who came up with a description of the universe that had no boundary – and no starting point. Whilst attending a conference at the Vatican in 1981, he apparently had no wish to counter the Pope's request that cosmologists stick to studying the universe after its creation – Hawking needed no creator.

A new way of formulating the equations of relativity came in 1986, at a quantum gravity workshop in Santa Barbara, California. Lee Smolin and Theodore Jacobson, later with Carlo Rovelli, hit upon a set of solutions to the equations based on 'quantum loops' in the gravitational field.

Quantum loops

The loops were the quanta of space. They did away with the need for a precise location, because it made no difference if the loops were displaced. Space is a knitted fabric of these loops, knotted and linked together. The loop concept appeared in other guises, during the development of quantum chromodynamics and in Roger Penrose's work explaining webs of particle interactions. In quantum gravity, these loop states become quanta of geometry. They are the smallest components of the universe – their size or energy is known as the Planck scale. Loop quantum gravity is a step towards an overarching theory, although there is still far to go. It does not say anything about the graviton, for instance. Other routes are still being explored, such as string theory.

Because of the huge energies needed to find a graviton, or any particle involved in the epoch when gravity broke away from the other forces, physicists can only dream of investigating quantum gravity at particle colliders. So there is no experimental evidence to support any of the models. The best bets in the interim are astronomical objects, especially black holes. Some black holes emit jets of particles – thought to be electron–positron pairs driven off when matter is sucked in. Around black holes gravity is very strong and unusual effects that violate relativity theory might be spotted.

Alternatively, the cosmic microwave background radiation is a hunting ground – its speckled hot and cold patches were produced by quantum variations in the young universe.

The condensed idea
Quanta of space

35 Hawking radiation

Black holes are pits in space-time that are so deep that not even light can escape. Except, that is, when quantum uncertainty allows. Stephen Hawking proposed that black holes can radiate particles – and information – causing them eventually to shrink.

By the 1970s, theories of quantum gravity were wallowing. Bryce DeWitt referred to his wavefunction of the universe as 'that damned equation' – no one knew what it meant. General relativists turned their attention to black holes. In the mid-1960s black holes were postulated to be the power source of the newly discovered quasars – galaxies whose cores were so bright they outshone all their stars.

The black hole idea was developed in the 18th century by the geologist John Michell and the mathematician Pierre-Simon Laplace. Later, after Einstein had proposed his relativity theories, Karl Schwarzschild worked out what a black hole would look like: a pit in space-time. In Einstein's theory of general relativity, space and time are linked and behave together like a vast rubber sheet. Gravity distorts the sheet according to an object's mass. A heavy planet rests in a dip in space-time, and its gravitational pull is equivalent to the force felt as you roll into the dip, perhaps warping your path or pulling you into orbit.

Event horizon

Black holes are so called because even light cannot escape their pull. If you throw a ball up in the air, it reaches a certain height and then falls back down. The faster you fling it, the higher it goes. If you hurled it fast enough it would escape Earth's gravity and whiz off into space. The speed that you need to reach to do this, called the 'escape velocity', is 11 km/s (or about 25,000 mph).

> God not only plays dice, but also sometimes throws them where they cannot be seen.
>
> Stephen Hawking, 1977

A rocket needs to attain this speed if it is to escape Earth. The escape velocity is lower if you are standing on the smaller Moon: 2.4 km/s would do. But if you were standing on a more massive planet then the escape velocity would rise. If that planet was heavy enough, the escape velocity could reach or exceed the speed of light.

If you pass far from a black hole, your path might curve towards it, but you needn't fall in. But if you pass too close to it, then you will spiral in. The same fate would befall a photon of light. The critical distance that borders these two outcomes is called the 'event horizon'. Anything that falls within the event horizon plummets into the black hole.

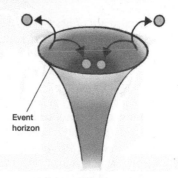

One member of particle-antiparticle pairs formed near the event horizon can escape the black hole's pull.

Frozen stars

Watching a piece of matter fall towards a black hole, you would see its progress stall. Time slows down near it. Light beams travelling in the vicinity of the black hole take longer to travel across the curved space-time landscape, and reach us.

As the matter crosses the event horizon, from a distant vantage point, time actually grinds to a halt. We see the material stop dead just at the point at which it falls in. In the 1930s Einstein and Schwarzschild predicted the existence of 'frozen stars', poised on the cusp of collapsing. The physicist John Wheeler renamed them 'black holes' in 1967.

The collapse of stars into black holes has been detailed by the astrophysicist Subrahmanyan Chandrasekhar. He showed that stars like our Sun are not heavy enough to be crushed by their

> The black holes of nature are the most perfect macroscopic objects there are in the universe: the only elements in their construction are our concepts of space and time.
> Subrahmanyan Chandrasekhar, 1983

own weight when their internal fusion engines switch off. Those more than 1.4 times the Sun's mass can collapse. But these would be propped up by quantum pressure because of the Pauli exclusion principle – forming white dwarfs and neutron stars. Only stars more than three times the Sun's mass can shrink further to produce black holes.

The existence of black holes in space was not discovered until the 1960s. Although they are dark, there are ways to tell if they are there. The intense gravity fields of black holes pull other objects, such as stars, towards them. And gas can also be drawn in, heating up and glowing as it encroaches.

A giant black hole has been located in the centre of our own galaxy. It has the mass of a million Suns squashed into a region just 10 million kilometres (30 light seconds) or so in radius. Astronomers have tracked the orbits of stars moving near the hole and seen them suddenly change course when they get very close. Just as comets have elongated orbits and are flung far out when they pass by the Sun, so these stars in the heart of the Milky Way move in strange ways around the black hole.

Black holes are the central engines in quasars. Gas falling into the black hole becomes superheated and glows fiercely. Stellar-mass black holes can also be identified by detecting the X-rays from hot gas swirling around them.

Evaporating black holes

Even if they are not swathed in gas, black holes are not completely black. Quantum effects mean that there is a chance that some radiation can escape, as Stephen Hawking ascertained in the 1970s.

Stephen Hawking (1942–2018)

Born during the Second World War, Stephen Hawking was raised in Oxford and St Albans, England. Hawking pursued physics at Oxford University and then moved to Cambridge to work with Dennis Sciama on cosmology. He held Isaac Newton's Lucasian professorship in mathematics there from 1979 to 2009. Diagnosed with ALS, a form of motor neuron disease, just after his 21st birthday, Hawking confounded his doctors and was as famous for his wheelchair-bound persona and computerized voice as for his science. Hawking's ideas include radiation from black holes and a boundary-free theory of the universe.

Spaghettification

Falling into a black hole has been described as being 'spaghettified'. Because the sides are so steep, there is a very strong gravity gradient within the black hole. If you were to fall into one feet-first, then your feet would be pulled more than your head and so your body would be stretched like being on a rack. Add to that any spinning motion and you would be pulled out like chewing gum into a scramble of spaghetti.

Particles and antiparticles are continually being created and destroyed according to Heisenberg's uncertainty principle in the vacuum of space. Should they pop into being near a black hole's event horizon, it is possible that one falls in and the other escapes. This escaping radiation is known as Hawking radiation. Because they lose energy as they radiate particles, black holes slowly shrink. Over billions of years they could evaporate away entirely.

There is more to the tale. If an object falls into a black hole, what happens to all the information contained in it? Is it lost for ever or are some of its quantum properties preserved and released in the Hawking radiation? If one of an entangled pair of particles fell in, would its partner know? Hawking believed that quantum information was destroyed. Other physicists disagreed vociferously. A famous bet was made. In 1997 John Preskill bet Hawking and Kip Thorne that information was not lost in black holes. In 2004, Hawking published a paper that claimed to resolve the paradox – showing that quantum effects at the event horizon could allow information to escape from the black hole. He sent Preskill an encyclopedia 'from which information can be retrieved at will'. Thorne, however, remains unconvinced and hasn't given way on his side of the bet.

The condensed idea
Not-so-black holes

36 Quantum cosmology

The universe's energetic and compact origins mean that quantum effects have left a mark on its grand-scale properties. Mysterious dark matter and dark energy may result from exotic particles and quantum fluctuations in the vacuum of space; and cosmic inflation might also have a quantum basis.

Winding back time, the universe must have been smaller and denser in the past. Some 14 billion years ago everything in it would have been crushed into a point. Its explosion from this moment was named the 'Big Bang' – originally in ridicule – by the British astronomer Fred Hoyle in 1949.

The temperature of the universe within a second of the Big Bang was so great that atoms were unstable and only their constituent particles existed in a quantum soup. A minute later, quarks pulled together to form protons and neutrons. Within three minutes, protons and neutrons combined according to their relative numbers to produce hydrogen, helium and traces of deuterium (heavy hydrogen), lithium and beryllium nuclei. Stars later furnished heavier elements.

Microwave background

Another pillar supporting the Big Bang idea was the discovery in 1965 of the faint echo of the fireball – the cosmic microwave background. Arno Penzias and Robert Wilson were working on a radio receiver at Bell Labs in New Jersey when they spotted an unexplained faint source of microwaves coming from all over the sky. The photons' origin was the hot young universe.

The existence of a faint microwave afterglow from the Big Bang, predicted in 1948 by Gamow, Alpher and Robert Hermann, originated in the epoch in which the first atoms formed, some 400,000 years after the fireball. At earlier times, the universe was filled with charged particles – protons and electrons flew around unattached. This plasma created an impenetrable fog, by scattering light photons. When atoms formed, the fog was cleared and the universe became transparent. From then on, light could travel freely across the universe. Although the young-universe fog was originally hot (some 3,000 kelvins), the universe's expansion has redshifted the glow from it so that we see it today at a temperature of less than 3 K (three degrees above absolute zero).

In the 1990s, NASA's COBE satellite mapped hot and cold patches in the microwave background, differing from the 3 K average temperature by 1 part in 100,000. This uniformity is surprising because when the universe was very young, distant regions of it could not communicate even at light speed. So it is puzzling that they nevertheless have almost the same temperature. These tiny variations in temperature are the fossil imprints of the quantum fluctuations in the young universe.

Deep connections

Three other properties of the universe also hint at deep connections forged in its earliest moments. First, light travels in direct lines across the vast reaches of space – otherwise distant stars and galaxies would be distorted.

Second, the universe looks roughly the same in all directions. This is unexpected. Having only existed for 14 billion years, the universe is more than 14 billion light years across in size (known as the 'horizon'). So light has not had time to get from one side of the universe to the other. How does one side of the universe know what the other side should look like?

Third, galaxies are sprinkled evenly across the sky. Again, this need not have been the case. Galaxies started out as just a

It is said that there's no such thing as a free lunch. But the universe is the ultimate free lunch.
Alan Guth

slightly overdense spot in the gas left over from the Big Bang. That spot started to collapse due to gravity, forming stars. The dense seeds of galaxies were set up by quantum effects, minuscule shifts in the energies of particles in the hot embryonic universe. But they could well have amplified to make large galaxy patches, like on a Friesian cow, unlike the wide scattering that we see. There are many molehills in the galaxy distribution rather than a few giant mountain ranges.

The three problems – flatness, horizon and smoothness – can be solved if the very early universe lay within its horizon. Then all its points could once have been in contact, setting its properties thereafter. If that was true, then some time later the universe must have suddenly become bloated, growing rapidly beyond its horizon, into the sprawling cosmos we see today. This rapid burst of expansion is known as 'inflation', and was proposed in 1981 by the American physicist Alan Guth. The slight density fluctuations, imprinted earlier by quantum graininess, became stretched and smeared out, making the universe smooth on large scales.

Dark side

Quantum effects might have other impacts on the universe. Ninety per cent of the matter in the universe does not glow but is dark. Dark matter is detectable by its gravitational effect but hardly interacts with light waves or matter. Scientists think it may be in the form of MAssive Compact Halo Objects (MACHOs), failed stars and gaseous planets, or Weakly Interacting Massive Particles (WIMPs), exotic subatomic particles, such as neutrinos and supersymmetric particles.

Today we know that only about 4% of the universe's matter is made up of baryons (normal matter comprising protons and neutrons). Another 23% is exotic dark matter. We do know that this isn't made up of baryons. It is harder to say what it is made from, but it could be particles such as WIMPs. The rest of the universe's energy budget consists of another thing entirely, dark energy.

> For 70 years, we've been trying to measure the rate at which the universe slows down. We finally do it, and we find out it's speeding up.
>
> Michael S. Turner, 2001

Albert Einstein came up with the dark energy concept as a way to compensate for the attractive force of gravity. With gravity alone, everything in the universe would collapse to a point. So some repellent force must counterbalance it. At the time, he didn't know that the universe was expanding and so believed it was static. He added a sort of 'anti-gravity' term to his equations of general relativity. But he quickly regretted it. Just as gravity could make everything collapse, so this anti-gravity could cause regions of space to rip apart. Einstein shrugged and imagined that this term wasn't needed – no one had seen any evidence for a repelling force. He kept the term in his equations, but he set it to zero.

That changed in the 1990s, when two groups found that distant supernovae were dimmer than they should be. The only explanation was that they were farther away than we thought. The intervening space must have stretched. Einstein's term came back into play – this negative energy term has been named 'dark energy'.

Anti-gravity

We don't know much about dark energy. It is a form of energy stored within the vacuum of free space that causes a negative pressure in empty regions. In places where matter is abundant – such as near galaxy groups and clusters – gravity soon balances or overwhelms it.

Because it is so elusive it is hard to predict how its presence will affect the universe in the long run. As the universe is pulled apart, galaxies will lose their grip and become spread more sparsely. Then dark energy can start to claw at their constituent stars. Once those stars die, the universe will become dark. Ultimately it would be a sea of scattered atoms and subatomic particles. Quantum physics would again rule.

The condensed idea
Early connections

37 String theory

In a modern twist on wave–particle duality, string theory attempts to describe elementary particles as wave harmonics on a vibrating string. The ultimate goal is to combine quantum physics and relativity and explain all four fundamental forces in one conceptual framework.

String theory is a parallel branch of physics that is developing an ambitious yet unique mathematical method for describing quantum and gravitational processes in terms of waves on multidimensional strings, rather than solid entities. It arose in the 1920s, when Theodor Kaluza and Oskar Klein hit upon using harmonics, like a musical scale, to describe some quantized properties of particles.

In the 1940s, it was clear that matter particles such as the electron or proton are not indefinitely small but have some extent. To explain why an electron possesses its own magnetism it must be a smeared-out spinning ball of charge. Werner Heisenberg wondered if this was because space and time broke down on extremely small scales. On larger scales, the fact that particles had reproducible behaviour in experiments meant that their quantum state held true, no matter what was going on below the surface. Building upon his matrix-mechanics description of the hydrogen atom, Heisenberg linked a particle's behaviour before and after some interaction using a matrix, or table of coefficients.

But quantum field theory was starting to show that particle processes didn't proceed in one grand step, but involved many tiny incremental ones. Heisenberg would have to supply a whole host of matrices to explain anything other than the simplest case. He tried to restructure his matrix notation, but to no avail.

In the 1960s, attention turned to finding ways of describing the strong nuclear force. Murray Gell-Mann was working on his quark theory of nucleons. Other theorists toyed with alternative mathematical pictures.

In 1970, Yoichiro Nambu, Holger Bech Nielsen and Leonard Susskind represented nuclear forces as vibrating one-dimensional strings. However, their model didn't take off and quantum chromodynamics superseded it. In 1974, John Schwarz, Joel Scherk and Tamiaki Yoneya extended the string idea to depict bosons. That they managed to include the graviton showed that string theory held promise for unifying all the forces.

Vibrating strings

Strings, like springs or elastic bands, want to contract to minimize their energy. This tension causes them to oscillate. Quantum mechanics dictates the 'notes' that they play, with each vibration state corresponding to a different particle. Strings may be open – with two end points – or closed, forming a loop.

The first string models weren't successful as they could only describe bosons. Building on the concept of supersymmetry, theories that included fermions – called superstring theories – became possible. A series of barriers were broken between 1984 and 1986, in what is known as the first superstring revolution. Realizing that string theory was capable of handling all the known particles and forces, hundreds of theorists joined the bandwagon.

The second superstring revolution came in the 1990s. Edward Witten drew together all the various superstring theories into one big 11-dimensional theory called M-theory (where 'M' means different things to different people, such as membrane or mystery). A flurry of papers followed between 1994 and 1997.

> There must be no barriers to freedom of inquiry. There is no place for dogma in science. The scientist is free, and must be free to ask any question, to doubt any assertion, to seek for any evidence, to correct any errors.
> J. Robert Oppenheimer, 1949

Since then, string theory has steadily progressed, shoring up its cathedral-like edifice as new experimental findings flow. But there is still no definitive theory – people say that there are as many string theories as there are string theorists. And string theory is arguably not yet in a fit state to be testable through experiments, making it something of a luxury as far as science goes.

The only way to truly test a physics theory, according to philosopher Karl Popper, is to prove a statement false. With no novel predictions that would prove string theory above other standard physics ideas, it is seen as something charming yet impractical. String theorists hope that one day that will change. Perhaps the next generation of particle accelerators will probe new physics regimes. Or perhaps research into effects such as quantum entanglement will advance such that hidden dimensions will be needed to explain them.

Theory of everything

String theory's ultimate goal is to describe a 'theory of everything', uniting the four fundamental forces (electromagnetism, the strong and weak nuclear forces and gravity) in one consistent picture. It's an ambitious goal and very far from being realized.

It's true that the rest of physics is fragmented. The Standard Model of particle physics has great power but its formulation was largely ad hoc, based on faith in underlying mathematical symmetries. Quantum field theories are an impressive feat, but their extrusion to include gravity is beyond a challenge. Those cancelled infinities – fixed by the mathematical trick of renormalization – still haunt quantum and particle theories.

Einstein's failure in the 1940s to unify quantum theory and relativity troubled him for the rest of his life. His peers thought him crazy for even attempting it. But the likelihood of failure hasn't put string theorists off pursuing their abstract quest. Will it be futile? Does it matter if a few scientists waste their time? Will we learn something along the way? Some physicists argue that string theory isn't real science. But not everything has to be. Pure mathematics helped Werner Heisenberg develop his matrix mechanics and let Murray Gell-Mann envisage quarks, after all.

What scope must a theory of everything have? Is it enough just to describe the physical forces? Or must it go further and include aspects of the world such as life and consciousness? Even if you describe an electron as a vibrating string, that may not tell you much about molecular bonds in chemistry, or how living cells are assembled.

Scientists fall into camps when it comes to such 'reductionism'. Some believe that we can create a 'bottom-up' picture of the world, constructed from matter and forces. Others argue that this is ridiculous – the world is so complex that a host of behaviours emerge from interactions that we have never thought of.

> I don't like that they're not calculating anything. I don't like that they don't check their ideas. I don't like that for anything that disagrees with an experiment, they cook up an explanation – a fix-up to say, 'Well, it still might be true'.
>
> Richard Feynman

Counter-intuitive aspects such as quantum entanglement and chaos make the world even harder to predict. The physicist Steven Weinberg believes that the building-block view is 'chilling and impersonal'. We must accept the world the way it is.

The condensed idea
Cosmic chimes

38 Many worlds

The Copenhagen interpretation's need for wavefunctions to collapse when a measurement is made troubled physicists who thought it unrealistic. Hugh Everett III found a way around it in the 1950s when he proposed that separate universes split off as quantum events unfold.

As scientists' understanding of particles and forces grew in the 1950s and 60s, so did their need to get more of a grip on what was really going on at the subatomic scale. Decades after it was put forward, the Copenhagen interpretation still reigned supreme – with its insistence that particles and waves are two sides of the same coin, described by a wavefunction, whose collapse is triggered when a measurement is taken.

The Danish physicist Niels Bohr's concept explained quantum experiments well, including the interference and particulate behaviour of light. Nevertheless, wavefunctions were hard to comprehend. Bohr thought them real. Others took them as mathematical shorthand for what was really going on. The wavefunction says with what chance an electron, say, is in some place or has some energy.

Worse, the Copenhagen interpretation puts all power in the hands of an 'observer'. When Schrödinger's cat sits poised in its closed box, radioactive danger untold, Bohr's supposition was that the feline is in a superposition of states – both alive and dead at the same time. Only when the box is opened is its fate sealed. But why should the cat care whether a human had seen it or not? Who observes the universe to ensure our existence?

Multiple universes

In 1957, Hugh Everett proposed an alternative view. He disliked the idea that wavefunctions must collapse when we make a measurement and that observers are needed to do it. How would a distant star know to exist if no person had seen it? He argued that everything in the universe exists at any moment in one state – the cat is really alive or it is dead. But to cope with all the possibilities, there must be many parallel universes where the alternative outcomes are realized. This is known today as his 'many worlds' theory.

While not all physicists believe in it, the many-worlds theory has proved popular with some. The American relativist Bryce DeWitt, who coined the 'many worlds' name, promoted the idea in the 1960s and 70s. Today many physicists use the 'multiverse' concept to explain otherwise inexplicable coincidences in cosmology, such as why the forces have the strengths they do, allowing atoms and life to exist.

Before Everett's proposal, the universe was thought to have a single path of history. Events unfolded with increasing time, producing a cascade of changes that fulfilled rules such as the second law of thermodynamics. In the many-worlds picture, every time that a quantum event happens, a new daughter universe splits off. The many universes – perhaps infinitely many of them – build up a branching structure, like a tree.

Although there is no bulk communication between each of the universes – they are separate and each goes its own way thereafter – some physicists have suggested that there might be a little meddling between split-off worlds. Perhaps those interactions explain interference experiments, or might even make time travel feasible between them.

Hugh Everett III (1930–82)

Hugh Everett was born and raised in Washington, DC. He studied chemical engineering at the Catholic University of America, taking a year out to visit his father, who was stationed in West Germany just after the Second World War.

Everett moved to Princeton University for his PhD, shifting from game theory to quantum mechanics. He was considered smart but too engrossed in science fiction books. In 1956 He went to work for the Pentagon on nuclear weapons modelling. At John Wheeler's request, Everett visited Niels Bohr in Copenhagen in 1959, but his work got a poor reception. Everett found the visit 'hell' and returned to his computing career.

In 1970 Everett's idea was popularized in an article by Bryce DeWitt, which drew a lot of attention. A follow-up book sold out in 1973. Science fiction writers loved the many-worlds concept. Everett died young, aged 51.

Benefits

The beauty of the many-worlds theory is that it avoids the need for wavefunction collapse and does away with the need for an observer to cause it. If Schrödinger's boxed cat is in a blend of possible states, then the experimenter must be also. The scientist who sees that the cat is alive is superposed with the scientist who will find him dead. Everett's concept thus solves many of the paradoxes of quantum physics. Everything that might have happened, already has in one universe or possibly could in another.

> I don't demand that a theory correspond to reality because I don't know what it is. Reality is not a quality you can test with litmus paper.
> Stephen Hawking

The universe can exist whether or not there's life. Schrödinger's cat is alive in one place and dead in another, not a mix of both. Wave–particle duality also makes sense as both eventualities are accommodated.

Everett worked out his model while he was still a graduate student, publishing it in his PhD thesis. The many-worlds idea wasn't taken up immediately and even drew scorn from some colleagues. Everett left research and went into defence work and computing. It took a popular article by Bryce DeWitt in *Physics Today* to draw it to wider attention in 1970.

Problems

Today the many-worlds concept has a mixed response. Its fans praise it for satisfying Occam's razor and doing away with much non-intuitive quantum behaviour. But is questionable whether many worlds is a testable theory. That depends to what degree the various universes interact and whether experiments can be proposed that prove that other universes exist. The jury is still out.

Those less taken with the interpretation argue that the splitting-off of universes is arbitrary – it's not really clear what these mean or how it happens. Everett's observerless picture gives no significance to the act of measurement, so it's not clear why, how or exactly when universes should branch off.

Other great puzzles of fundamental physics also remain unexplained – such as the direction of time and why entropy increases according to the second law of thermodynamics. It's not clear whether quantum information can travel across the universe faster than light – whether the entire universe splits off every time a particle pops into existence around a black hole on the far side of the universe, for instance. Some of the parallel universes couldn't exist if their physical properties were incompatible.

> One should put one's trust in a mathematical scheme, even if the scheme does not appear at first sight to be connected with physics.
> Paul Dirac, 1977

Stephen Hawking is one critic who views many-worlds theory as 'trivially true', a useful approximation for calculating probabilities rather than a deep insight into the real universe. Dismissive of attempts to even try to understand the meaning of the deep quantum world, he has said, 'When I hear of Schrödinger's cat, I reach for my gun.'

The condensed idea
Parallel universes

39 Hidden variables

The fact that the quantum world could only be described in terms of probability worried some physicists, including Albert Einstein. How could cause and effect be explained if everything happens by chance? A way around this is to assume that quantum systems are wholly defined, but that there are hidden variables we have yet to learn.

Albert Einstein famously didn't like the Copenhagen interpretation of quantum mechanics, declaring that 'God does not play dice'. What concerned him was that probabilistic treatments of quantum mechanics were not deterministic – they could not predict how a system would evolve in the future from a particular state.

If you know the properties of a particle now then, due to Heisenberg's uncertainty principle, you could not also know them some time later. But if the future depends on chance occurrences, why is the universe ordered and driven by physical laws?

As Einstein, with Boris Podolsky and Nathan Rosen, encapsulated in 1935 in the EPR paradox, quantum mechanics must be incomplete. Because messages can't travel faster than the speed of light, twin particles with entangled quantum rules that fly apart must always 'know' what state they are in. An observation of the state of one particle tells us something about the other, but not because any wavefunction is collapsing. The information was inherent to each particle and contained in 'hidden variables', Einstein reasoned. There must be a deeper level of understanding that is hidden from our view.

Determinism

In the 1920s and 30s, physicists puzzled over the meaning of quantum mechanics. Erwin Schrödinger, who had proposed his wave equation in 1926, believed that the wavefunctions that described a quantum system were real physical entities. His colleague Max Born struggled more to comprehend the picture. In a paper Born noted that the probabilistic interpretation of the wave equation had implications for determinism – cause and effect.

Born considered that further atomic properties might one day be discovered to explain the consequences of a quantum event, such as a

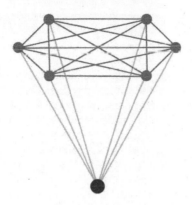

Bohm imagined that a particle possesses a web of
'hidden knowledge' of all the physical properties it could have,
but quantum mechanics limits what we can know of them.

collision between two particles. But in the end he backed the
wavefunction approach and accepted that not everything was
knowable: 'I myself am inclined to give up determinism in the world
of atoms. But that is a philosophical question for which physical
arguments alone are not decisive.'

Einstein too was suspicious of wavefunctions. He saw Schrödinger's
equation as only describing atoms in a statistical sense, not completely,
although he couldn't prove it. 'Quantum mechanics is very worthy
of regard. But an inner voice tells me that this not yet the right track,'
he remarked.

At a conference in Belgium in 1927 the French physicist Louis de
Broglie presented a hidden-variable theory that maintained
determinism. A 'pilot wave' guided each particle through space, he
suggested. Einstein had also considered this possibility, but he had
cast the idea aside, and stayed silent. Other physicists also steered
clear. The majority were swayed by the confidence of Born and Werner
Heisenberg, who bravely announced that quantum mechanics was
now a complete theory. The indeterminism was real, within the
domain of experiments to which it applied, they believed.

After Niels Bohr proposed his Copenhagen interpretation of
quantum mechanics – requiring an observer to collapse the
wavefunction during measurement – in 1927 he and Einstein debated

fiercely whether it made sense. Einstein's best challenge was the EPR paradox, which raised the possibility of an observer on one side of the universe collapsing a wavefunction on the other side instantaneously, in violation of special relativity.

Guiding waves

In 1952 David Bohm resurrected hidden-variable theory, when he unwittingly rediscovered de Broglie's unpublished 'guiding wave' idea. Bohm believed that particles such as electrons, protons and photons are real. We can see individual photons build up on a detector, for instance, or electrons create pulses of charge on hitting an electric plate. So, he reasoned, Schrödinger's wavefunction must be describing the probability of it being somewhere.

To guide the particle as to where it should be, Bohm defined a 'quantum potential'. It hosts all the quantum variables and responds to other quantum systems and effects and is linked to the wavefunction. So the position and trajectory of a particle is always defined, but because we don't know all the properties of the particle at the outset we must use the wavefunction to describe the probability that the particle is somewhere or has some state. The hidden variables are the particle positions, not the quantum potential or wavefunction.

Bohm's theory retains cause and effect – the particle is travelling along some path just as in classical physics. It eliminates the need to collapse a wavefunction. But it doesn't get around the EPR paradox or 'spooky' action at a distance. If you change a detector then the particle's wave field also changes instantaneously. Because it acts irrespective of distance, the theory is said to be 'non-local'. So it also violates special relativity, leading Einstein to call the theory 'too cheap'.

> In some sense man is a microcosm of the universe: therefore what man is, is a clue to the universe. We are enfolded in the universe.
> David Bohm

Bohm showed that it was possible to write down a hidden-variable version of quantum mechanics. The next steps were to test it. In 1964, John Bell derived a set of imaginary experiments whose outcomes could be consistent with hidden-variable theory. If the results differed from those predictions, then quantum entanglement would be true. In the 1980s physicists managed to perform those tests. They ruled out the simplest case of 'local' hidden variables, where information is constrained by the speed of light. Instantaneous long-distance correlations or quantum entanglement are indeed needed.

The condensed idea
Known unknowns

40 Bell's inequalities

In 1964 John Bell encapsulated the difference between quantum and hidden-variable theories in equations. He proved that correlations between particles should behave differently if they are imprinted at birth or upon measurement.

Quantum mechanics is troubling. Its probability-based assertions and fundamental uncertainties, even about basic properties such as energy and time, position and momentum, seem to defy explanation.

Adherents to the 1927 Copenhagen interpretation, including Niels Bohr and Erwin Schrödinger, accepted the fact that there is a limit to what we can know about the subatomic world. Particles such as electrons also behave like waves, and the only way to describe what we do know about them is in a mathematical form, as a wavefunction.

Albert Einstein and Louis de Broglie in the 1930s, and later David Bohm in the 1950s, clung to the belief that electrons and photons and other particles are real entities. They exist – it's just that we can't know all about them. Quantum mechanics must be incomplete. A set of 'hidden variables' could explain some of its more counter-intuitive aspects.

The EPR paradox defied explanation. The properties of two correlated particles sent flying off in opposite directions across the universe must remain linked, even if they become so distant that a light signal from one cannot reach the other. This reasoning predicts 'spooky' action at a distance.

It now seems that the non-locality is deeply rooted in quantum mechanics itself and will persist in any completion.
John Bell, 1966

Just as electrons are limited in how they fill their orbitals, quantum rules tie together particles. If one particle (say a hydrogen molecule) splits in half, then conservation principles mean that the spins of both resulting particles will be opposite. If we measure one particle's spin as 'up', then we immediately know that the spin of the other must be 'down'. In quantum terms, the wavefunction of the second particle collapses at exactly the same time as the first, no matter how far apart the particles are.

Einstein and his colleagues worried that this is unreasonable. No signal can travel faster than light, so how would the measurement of the first particle be conveyed to the second? Einstein's reasoning relied on two assumptions: locality, that nothing can travel faster than light, and realism, that particles exist whether they are 'observed' or not. Einstein's thinking was in terms of 'local realism'.

Bell's theorem

In 1964 John Bell took this thinking further. If hidden variables and local realism were both true, then any decision made about a measurement on one nearby particle would not affect the property of the distant one. If the remote particle already knew what state it was in, then it shouldn't care whether or not you decided to measure the particle at hand using interference or scattering.

Bell defined specific cases where this behaviour would clash with the wilder predictions of quantum mechanics. He defined quantities that could be measured to test this, such that if a value greater or less than some limit was obtained the evidence would point to quantum mechanics or hidden variables. These mathematical statements are known as 'Bell's inequalities'.

John Stewart Bell (1928–90)

John Bell was born in Belfast, Northern Ireland, and studied physics at Queen's University, Belfast. He completed his PhD in nuclear and quantum physics at the University of Birmingham in 1956. Bell worked with the UK Atomic Energy Research Establishment, near Harwell, Oxfordshire, and then moved to the European Council for Nuclear Research (CERN, Conseil Européen pour la Recherche Nucléaire), in Geneva, Switzerland. Here he worked on theoretical particle physics and accelerator design, but found time to investigate the foundations of quantum theory. In 1964, after spending a sabbatical year working in the USA, Bell wrote a paper entitled 'On the Einstein–Podolsky–Rosen Paradox', in which he derived the Bell theorem in terms of an expression violated by quantum theory.

Modifying the EPR example, Bell imagined two fermions whose spin was complementary, such as two electrons, one with spin up and the other spin down. Their properties were correlated, perhaps because they both started out as a single particle that decayed. The two particles travel off in opposite directions.

It is not known which one has which value of spin. Measurements of both are made, at their respective locations. Each observation would yield a result of spin up and spin down. Each measurement is carried out independently, without knowing anything about the other.

> Nobody knows just where the boundary between the classical and the quantum domain is situated . . . More plausible to me is that we will find that there is no boundary.
>
> John Bell, 1984

The probability of measuring a particular direction of spin depends on the angle with which it is measured, from 0 to 180 degrees. The chance is +1 if you measure it in exactly the same direction as the spin axis; it is −1 if measured in the opposite direction; and half that if measured at a direction perpendicular. At angles in between, the different theories predict different chances of measurements.

Bell's theorem gives the statistics of what would be seen for many trials of the experiment measured at various angles. For hidden-variable theory there is a linear relationship between these points. For quantum mechanics, the correlation varies like the cosine of the angle. So by making measurements in many different directions it is possible to tell which is happening.

Bell concluded: 'there must be a mechanism whereby the setting of one measuring device can influence the reading of another instrument, however remote. Moreover, the signal involved must propagate instantaneously.'

Predictions tested

It took more than a decade for experiments to be good enough to really test Bell's predictions. In the 1970s and 80s a series of them proved quantum mechanics is correct. They rule out local hidden-variable theories, ones whose quantum messaging is limited by the speed of light. And so they prove that faster-than-light signalling is happening on the quantum scale. Some variants of hidden-variable

theories are still possible, if they are non-local, or open to
superluminal signalling. Bell welcomed the discovery but was also
disappointed: 'For me, it is so reasonable to assume that the photons
in those experiments carry with them programs, which have been
correlated in advance, telling them how to behave.' It was a pity that
Einstein's idea didn't work.

Bell's theory is one of the most important in fundamental physics.
It doesn't quite prove quantum mechanics exactly – some loopholes in
its reasoning have been identified. But it has held up over numerous
attempts to disprove it.

The condensed idea
Quantum limits

41 Aspect experiments

Experiments to test Bell's inequalities in the 1970s and 1980s proved that quantum entanglement does happen. Twin particles both seem to know when one of the pair is observed, even if one is extremely distant. As a result, quantum information is not stored once and for all but is cross-linked and responsive.

In 1964, John Bell wrote down a series of mathematical statements that would hold if the hidden-variables picture of quantum physics was correct – and particles carried a full portfolio of parameters with them. If these rules were violated then the weirder aspects of quantum mechanics must be true. Spooky action at a distance, faster-than-light messaging and quantum entanglements indeed take place.

It took more than a decade to devise definitive experimental tests of Bell's inequalities. The reason is that it is hard to do. First you need to identify an atomic transition that gives off pairs of matched particles, a property of each particle that depends on orientation and can be measured reliably and accurately, and an experimental design for doing so.

In 1969 John Clauser, Michael Horne, Abner Shimony and Richard Holt suggested using as the entangled particles photon pairs produced by excited calcium atoms. By raising the energy of the outer electron pair in calcium into higher orbitals and letting them fall back, two photons would be emitted. Because they obeyed linked quantum rules, the pair would have correlated polarizations, a characteristic that had been known since the late 1940s.

In 1972 Clauser and Stuart Freedman performed the very first successful experiment to test Bell's inequality. It was difficult to excite and capture the paired photons and took 200 hours of running time. The photons' polarizations had to be detected in the blue and green parts of the spectrum, but detectors at the time weren't very sensitive. In the end, the result agreed with the quantum mechanics prediction. But Clauser and his colleagues had to apply a statistical fudge for dealing with the missing photons, so it wasn't the end of the story.

Further experiments, on calcium as well as on excited mercury atoms and using photon pairs produced in positron annihilations, followed. Most also backed quantum mechanics, although some were inconclusive. The accuracy of the experiments improved with detector technology and as lasers were introduced, making it more efficient to excite atoms so that more photons could be given off.

Aspect's tests

In the late 1970s the French physicist Alain Aspect upgraded the experiment. Again using vaporized calcium, he tuned two lasers to the precise frequencies needed to make the outer electron pair quantum-leap into higher shells and be released in a cascade. He monitored beams of light coming off in two opposite directions, one tuned to each photon frequency, green and blue.

Because the time between each photon pair being given off was longer than the interval between each photon's release within the pair, the beams simultaneously measured the correlated pairs. Moreover, any communication between the separated photons would have to travel at twice the speed of light to connect them.

Just as Polaroid sunglasses reduce glare by blocking reflected light, the polarization of the photons in each beam was measured using special prisms. The prisms transmitted vertically polarized light well (around 95% of the light went through) but almost all horizontally polarized light (again around 95%) was blocked and reflected. By rotating the prisms, Aspect's team could measure how much light of intermediate polarizations came through.

Aspect, Philippe Grangier and Gérard Roger published their results in 1982. They were consistent with a cosine variation in polarization with angle, in support of quantum mechanics. Local hidden variables predicted a linear fall-off. Their result had a much higher statistical significance than previous efforts and was a landmark.

> The most exciting phrase to hear in science, the one that heralds new discoveries, is not 'Eureka!' (I found it!) but 'That's funny . . .'
>
> Isaac Asimov

As a consequence of this, local hidden-variable theories were dead, or certainly on the critical list. There was still a little room for exotic types of hidden variables that could switch faster than light speed, but simple models that relied on direct communication at light speed or less were ruled out. So measurement of one particle did affect the other, no matter how far away. Quantum states are indeed entangled.

Plugging loopholes

Critics complained that the experimental tests were not perfect and had loopholes. The detection loophole was one, plugged in the analysis by Clauser: not every photon will be detected so a statistical way of accounting for this is needed. A second issue is the communication loophole – that one detector might somehow pass information to the other owing to the limited size of the experiment. This could be excluded by switching the apparatus faster than any message could be sent.

Aspect had aimed to avoid this loophole by using an opposing twin-beam set-up in his first experiment. But to make sure, he changed the polarizer settings while the photons were in flight. His further experiment again proved that quantum theory holds. In 1998, an Austrian team led by Anton Zeilinger went further, by making the choice of detector very fast and random. There was no way one side of the experiment could know what the other side was doing. Again quantum mechanics was backed. Then in 2001 teams of American physicists sealed the remaining 'fair sampling' loophole by capturing every correlated photon from a beryllium-based experiment. The findings are now unequivocal. Quantum information is entangled.

Distant entanglement

Today, physicists have shown that entanglement can be maintained over vast distances. In 1998, Wolfgang Tittel, Jürgen Brendel, Hugo Zbinden and Nicolas Gisin at the University of Geneva managed to measure entanglement between pairs of photons across a distance of 10 km. The photons had been sent through fibre-optic cables through tunnels across Geneva.

In 2007, Zeilinger's group communicated using entangled photons over 144 km, between the Canary Islands of La Palma and Tenerife. Entanglement is now being investigated for long-distance quantum communication.

Clauser's and Aspect's experiments, and the many others since, have now conclusively shown that local hidden-variable theories don't work. Quantum entanglement and faster-than-light communication does happen.

The condensed idea
Faster-than-light communication

42 Quantum eraser

Variants of Young's double-slit experiment give us some insight into wave–particle duality. Interference only arises when photons are correlated but their paths are uncertain. Once their trajectories are known, they act as particles and fringes disappear. This behaviour can be controlled by entangling or erasing quantum information.

Wave–particle duality lies at the heart of quantum physics. As Louis de Broglie proposed, everything has both wave and particle characteristics. But these two facets of nature cannot manifest at the same time. They appear under different circumstances.

In the 19th century Thomas Young showed with his double-slit experiment how light behaves like waves when passing through a gap, the criss-crossing trains producing interference stripes. In 1905 Albert Einstein showed that light also behaves as a stream of photons. Electrons and other elementary particles may also interfere in the right circumstances. The Danish physicist Niels Bohr imagined that waves and particles were two sides of the same coin. Werner Heisenberg explained that full knowledge about certain complementary properties such as position and momentum was mutually exclusive. Might such unpredictability underlie wave–particle duality too?

In 1965 Richard Feynman wondered what would happen if we could measure which slit a particle had travelled through in Young's experiment. As we fired electrons through twin slits, we could shine light upon the apparatus and so, by detecting the scattered light, tell the routes of individual electrons. If we know an electron's position, and thus treat it as a particle, the interference fringes should disappear, he reasoned.

In 1982, the theoretical physicists Marlan Scully and Kai Drühl imagined another experiment with two atoms acting as the light sources. If we use a laser to excite an electron in each atom into the same higher energy level, then each electron would drop back and release a similar photon. Both would have the same frequency, and so it would be impossible to tell from which atom they came. These photons should then interfere, giving fringes. But we can go back and find out from which atom a given photon has come. We could measure the energy of the remaining atoms – the one that has lost energy would be the host of the released photon. We can measure the atoms after the photon is released. So naively, we should be able to see both the wave and particle sides of the light at once.

Yet the Copenhagen interpretation tells us categorically that we cannot see both. According to quantum mechanics, we have to think of the whole system and its wavefunction. By observing the state of the atoms, even after the photon has fled, we affect the whole experiment. If we tell it so, the photon will act as a particle and the interference will disappear.

Erasure

What if we measure the atoms but don't look at the result? In theory, the interference fringes should remain if we don't know anything about the photons' paths. In reality, if we measure the energies of the remaining photons but keep that secret, the fringes don't come back.

One way of measuring the energies but filing away the information is to fire a further laser photon at them both. The atom that produced the first photon would again be excited; a new third photon would be given off. But we now couldn't tell which atom it came from – either one might have decayed.

However, this is not enough for the interference fringes to reappear. The interfering photons don't know anything about the third photon. It is necessary to correlate both groups before the fringes can arise. In the case above, we could erase the information contained in the third photon, while keeping it part of the whole system. By detecting the third photon in such a way that we can't tell which atom it came from, the quantum uncertainty is brought back. For instance, the third photon could be caught by a detector placed in between the two atoms. The chance of this happening would be 50%, so it would be uncertain. But its detection would reset the system such that we really wouldn't know anything about the interfering photons' paths. Such an experiment is known as a quantum eraser because it destroys quantum knowledge about a system.

> I don't feel frightened by not knowing things, by being lost in a mysterious universe without any purpose, which is the way it really is, so far as I can tell. It doesn't frighten me.
> Richard Feynman, 1981

If we look at a deeper level, an original interference photon becomes correlated with the third photon. There are two possibilities – that the third photon is detected or not. And each case has an interference pattern. However, both are shifted in phase, such that when combined they cancel out. So the appearance of the third photon – with its uncertainty intact – adds an interference pattern that cancels out the first. When its fate is learned, and it is detected, the system picks one set of fringes.

Entangled interference

In 1995, Anton Zeilinger's group in Innsbruck, Austria, made a similar observation, using entangled pairs of photons generated by laser excitation of a crystal. By using very low levels of red and infrared light they could essentially follow individual photons through their experiment. First they produced a beam of excited photons, and passed some of it back through the crystal to make a second beam. Interference was produced where they crossed. But if each beam was made distinguishable – so that the path of a given photon could be

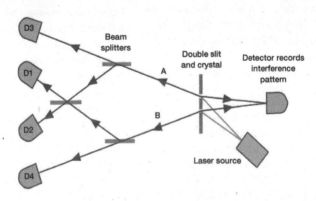

Light from each slit follows different paths, A and B, and is split again. Which-way information is erased for photons hitting D1 or D2, and not for D3 or D4.

ascertained – by changing its polarization, the fringes disappeared. The interference pattern didn't reappear until the two pathways were scrambled so that all positional information was lost.

Even stranger, it doesn't matter when the decision is made to apply the quantum eraser. You can even do so after the interfering photons are detected. In 2000 Yoon-Ho Kim, with Scully and colleagues, performed such a 'delayed choice' quantum eraser experiment. The interference pattern can be controlled by choosing whether or not to learn about a photon's trajectory after it has already hit the detector. The interference stripes only appear when the secondary uncertainty is resolved.

So there is a link between complementarity and non-local effects in quantum physics. Interference only works because of these entangled long-distance correlations. And it is simply impossible to measure both wave and particle properties at once.

The condensed idea
Ignorance is bliss

43 Quantum decoherence

Quantum systems easily become entangled with others, so that their wavefunctions combine. Whether they do so in phase or not dictates the result. Quantum information can thus easily leak out, leading to the loss of cohesion of a quantum state. Larger objects decohere faster than smaller ones.

In the quantum world, everything is uncertain. Particles and waves can't be distinguished. Wavefunctions collapse when we pin something down through measurement. In the classical world, everything seems more solid. A speck of soot is a speck of soot from one day to the next.

Where does the division between the quantum and classical worlds kick in? Louis de Broglie assigned a characteristic wavelength to every object in the universe. Big objects like footballs have a small wavelength, so their behaviour is particle-like. Tiny things like electrons have wavelengths closer to their size, so their wave-like properties are visible.

In the Copenhagen interpretation of quantum mechanics, Niels Bohr proposed that wavefunctions 'collapse' whenever a measurement is made. Some of their inherent probability is lost when we recognize a characteristic with certainty. It is irreversible. But what is going on when a wavefunction collapses or when we make a measurement? How are the fuzzy uncertainties converted into a hard outcome?

Hugh Everett bypassed the issue when he proposed the 'many worlds' concept in 1957. He treated the entire universe as having one wavefunction, which evolves but never collapses. An act of measurement is an interaction or entanglement between quantum systems, which spins off a new universe. Even so, Everett couldn't explain the exact point at which this happens.

In later 'guide wave' quantum theories, such as the one by de Broglie and David Bohm that sought to describe wave–particle duality in terms of a particle in a quantum potential, measurements distort the motion of the particle in its quantum field. It's a bit like bringing one mass up near another in general relativity – space-time shifts to blend the gravitational influences. There is no actual collapse of the particle's wavefunction, it just changes form.

Overcoming decoherence poses a significant challenge for quantum computers, which rely on quantum states being able to be stored for long periods of time. A measure called 'quantum discord' has been proposed to describe the degree of correlation between quantum states.

Decoherence

Today, the best explanation for the replacement of possibility with certainty is the concept of decoherence, noted in 1970 by the physicist Dieter Zeh. When two or more wavefunctions come up against one another, such as when a measuring apparatus is brought near a quantum entity, how they interact depends on their relative phases. Just as crossing light or water waves amplify or cancel out when they interfere, wavefunctions can be boosted or erased when they mix.

The more interactions a wavefunction has to contend with, the more scrambled it becomes. Eventually it decoheres and loses its wave-like aspects. Decoherence is much more significant for large objects – they lose quantum cohesion fastest. Small entities such as electrons retain their quantum integrity for longer. Schrödinger's cat, for instance, would pretty soon regain its feline form even if not observed, because its wavefunction would almost instantaneously degrade.

This is a comforting idea. It puts our familiar macroscopic world on a surer footing. But puzzles remain with the approach. For instance, why does decoherence act so uniformly over a quantum behemoth like a cat? Couldn't half of the animal stay in quantum penury, the other becoming real? Could it be literally half alive and half dead?

Also, what restricts the outcome of the decayed wavefunction to the appropriate observables? Why does a photon appear when needed, or a light wave when a slit is placed in the way? Decoherence tells us little about wave–particle duality.

Large systems

One way of learning more is to devise and study a macroscopic phenomenon or object that exhibits quantum behaviour. In 1996 and 1998 the French physicists Michel Brune, Serge Haroche, Jean-Michel Raimond and their colleagues manipulated electromagnetic fields into superposition of states using rubidium atoms and saw their quantum integrity decay. Other groups have tried to build bigger and better Schrödinger's-cat-like scenarios.

> There is one great difficulty with a good hypothesis. When it is completed and rounded, the corners smooth and the content cohesive and coherent, it is likely to become a thing in itself, a work of art.
> John Steinbeck, 1941

The quantum behaviour of large molecules is another avenue. In 1999 Anton Zeilinger's group in Austria managed to observe the diffraction of buckyballs – football-shaped cages of 60 carbon atoms called buckminsterfullerenes after the architect Buckminster Fuller. In terms of relative scale, the experiment was like firing a football at a goal-sized gap and seeing the ball interfere and behave like a wave. The buckyball's wavelength was one-400th the molecule's physical size. Another large system where decoherence effects can be studied is a superconducting magnet, which is often in the form of a supercooled metal ring, centimetres in diameter. Superconductors have unlimited conductivity – electrons can pass unhindered through the material.

The superconducting ring adopts particular energy levels, or quantum states. So it is possible to see how they interfere if you place them near to one another, say with currents flowing in opposite directions, clockwise and anticlockwise. A plethora of studies have now proved that systems decohere more quickly the larger they are.

Quantum leakage

Decoherence can be thought of as the leaking of quantum information into the environment, through many small interactions. It doesn't actually cause wavefunctions to collapse, but it mimics it as the quantum components of a system steadily become decoupled.

So decoherence doesn't solve the measurement problem. Because they must be large enough for us to read, measuring devices are simply complex quantum systems placed in the way of the pristine one that we are trying to observe. So the many particles making up the detector each interact with their quarry in complex ways. Those many entangled states gradually decohere, until we are left with a jumble of separate states. This quantum 'sandpile' becomes the result of the final measurement, with the extraneous quantum information from the original system sucked out.

All in all, this picture of the tangled web of quantum interactions shows that 'realism' is dead. Like 'localism', the transmission of signals through direct light-speed-limited communication, 'realism', the idea that a particle exists as a separate entity, is a charade. The apparent reality of the world is a mask put on to hide the fact that it is really made of quantum ashes.

The condensed idea
Leaky information

44 Qubits

Quantum computers could one day replace silicon-based technologies. They could be powerful enough to crack almost any code. Still only prototypes, they handle pieces of binary data in the form of 'quantum bits', or states of atoms. Based on quantum mechanics, they could exploit phenomena such as entanglement to make millions of calculations at once.

The tiny dimensions of quantum systems and their ability to exist in different states raise the possibility of building radically new types of computers. Rather than using electronic devices to store and process digital information, individual atoms become the heart of a quantum computer.

Proposed in the 1980s and developing rapidly in recent decades, quantum computers remain a long way from being realized. Physicists have only succeeded in linking tens of atoms in ways that can be used to perform calculations. The main reason is that it is hard to isolate the atoms – or other basic building blocks – so that their quantum states can be read yet remain immune to disturbances.

Conventional computers work by breaking down numbers and instructions into binary code – a series of 0s and 1s. Whereas we usually count in multiples of ten, computers think in factors of two: the numbers 2 and 6 would be expressed in binary notation as '10' (one 2 and zero 1s) and '110' (one 4, one 2, and zero 1s). Each 0 or 1 'binary digit' is known as a 'bit'. An electronic computer translates this binary code into physical states, such as on or off, within its hardware. Any either-or distinction would work as a means of storing binary data. The strings of binary numbers are then handled through banks of logic gates, hardwired into silicon chips.

Quantum bits

Quantum computers are qualitatively different. They are also based on on-off states – quantum bits or qubits for short – but with a twist. Like binary signals, qubits can take on two different states. But unlike normal bits they can also exist in a quantum blend of those states.

A single qubit can represent a quantum superposition of two states, 0 and 1. A pair of qubits may superpose four states, and three qubits

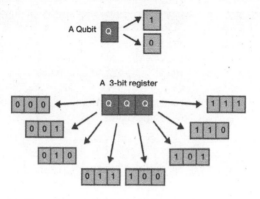

Three qubits can represent eight states simultaneously.

cover eight states. Each time you add another qubit, the number of states blended doubles. By contrast, a traditional computer would only be in one of these states at any one time. This rapid doubling of links between qubits gives the quantum computer its power.

Another benefit of the quantum world that can be harnessed for computation is entanglement. The behaviour of qubits far apart from one another may be tied by quantum rules. Flicking one into one state may simultaneously change that of another, bringing both speed and versatility to the mechanisms for solving mathematical problems.

For these reasons, quantum computers have the potential to be much faster than conventional machines for performing some types of calculations. Quantum networks are particularly efficient at and suited to solving problems that require rapid scaling or complex networks of linked communication.

In 1994 the field received a boost when the mathematician Peter Shor developed an efficient algorithm for factorizing large integers – working out the prime numbers that multiply to give that number – on a quantum computer. Shor's algorithm has now been implemented by several teams using handfuls of qubits. Although it's a great technical breakthrough, the results to date are admittedly far from mind-blowing: the number 15 has been shown to be 3 times 5, and 21 is 3 times 7. But these are very early days. When sizeable quantum computers become available, Shor's algorithm might unleash its power. It could potentially be used to break all current cryptography codes on the internet, necessitating different ways of securing information online.

Staying coherent

How can you build a quantum computer? First you need some qubits. Qubits can be assembled from more or less any quantum system that can adopt two different states. Photons are the simplest – perhaps using two distinct directions of polarization, vertical and horizontal. Atoms or ions with differing electron arrangements have been tried, as have superconductors, with electron currents flowing clockwise or anticlockwise.

Just as Schrödinger's cat is potentially both alive and dead when it lies unseen in its box, qubits superpose outcomes until their final state is locked down through a measurement. Also like that of the fabled cat, qubit wavefunctions are susceptible to partial collapse through many tiny interactions with the objects in their environment. Limiting this quantum decoherence is a major challenge for quantum computing. Within the device, it is important to keep the qubits isolated, so that their wavefunctions don't get disrupted. At the same time the qubits must be able to be manipulated.

Individual qubits, such as atoms or ions, may be embedded in tiny cells. A copper and glass casing can protect them from stray electromagnetic fields and allow electrodes to be connected. The atoms must be held in vacuum conditions, to avoid interactions with other atoms. Lasers and other optical devices can be used to alter the energies and quantum states of the qubits, such as their electron levels or spin. To date, only small prototype quantum 'registers', of tens of connected qubits, have been made. There are many challenges. First, even building one qubit and keeping it isolated is hard. Keeping it stable for long periods without losing quantum coherence is difficult, as is ensuring that it gives accurate and reproducible results – every time you multiply 3 by 5 you want to get the right answer. Joining several qubits together compounds the complexity. And as the qubit arrays grow larger, the difficulty of controlling the whole assembly escalates. The possibility of stray interactions rises and accuracy suffers.

Future computers

As the silicon-chip technology of traditional computers reaches its limit, we can look forward to quantum techniques that will deliver a whole new level of power. A quantum computer could simulate just about anything and could even be key to creating an artificially intelligent machine.

By performing so many calculations simultaneously, quantum computers are in effect doing mathematics across many parallel universes, rather than on parallel machines. Like Shor's function, new types of algorithms will be needed to exploit this power. But the source of a quantum computer's strength is its weakness. Because they are so sensitive to the environment they are fundamentally fragile.

The condensed idea
Truly parallel computing

45 Quantum cryptography

Our ability to send private encoded messages is under threat as computers become so powerful that they can crack most codes. A foolproof way is to employ quantum and entanglement to scramble messages. Any eavesdropper would alter the quantum system's state, making it obvious if there had been an intrusion and destroying the message itself.

Whenever you check your bank account online or send an email across the internet your computer exchanges messages in a scrambled format so that no one except the recipient can read them.

> I knew that the day on which I should be able to send full messages without wires or cables across the Atlantic was not far distant.
>
> Guglielmo Marconi

The letters and numbers are transformed into a coded message, which is then reassembled at the other end, using a key to translate it. Codes have long been used as a way to prevent people from prying. Roman emperor, Julius Caesar, used a simple cipher to pass on his messages: merely swapping one letter for another. Replacing each letter with one two places to the right in the alphabet would turn the message 'HELLO' into the inscrutable 'JGNNQ'.

In the Second World War the Nazis built machines to automate the coding process for their secret communications. The most sophisticated device was called Enigma. The beauty of using a machine to encode sentences was that the precise matching of the letters from the original to the encrypted version depended on how each machine was built. There was no simple rule an interceptor could follow – you had to have a matching machine to unravel the code.

British mathematicians working at Bletchley Park, the government's secret code-breaking establishment, including Alan Turing, famously managed to beat Enigma by working out the probabilities of certain combinations of letters occurring. For the 'HELLO' message above Turing might have spotted that the double 'NN' is likely to be 'LL' or maybe 'EE' or 'OO'. With enough words, the code could be cracked. The German notes he deciphered at Bletchley turned the course of the war in favour of the Allies.

Secret keys

As communications technologies have advanced, more and more complicated codes have been needed. Even machine-based ones are not immune. For a shatterproof code you ideally need a one-off and random mapping from one letter to the coded one. If the reader has the same key to the code then they can translate the message.

Keys are often used in one of two ways, known as public- and secret-key cryptology. In the first case, the sender picks two linked keys. One she keeps to herself, the other she makes public. Just like passing mail through a metal mailbox with two doors, anyone can send mail to her using the partial code for the public key. But only she has the second key with which to completely decode it. The second method uses one key, which must be shared between two people wishing to have an exchange. In this case the code is only secure as long as the key is kept secret.

Neither method is fail-safe. But some quantum tricks can shore them up. Publicly shared keys must be enormously long to thwart systematic attempts to crack them. But this slows down the encryption and deciphering processes. The faster computers get, the longer the keys must become. When quantum computers become feasible, most public-key codes could be cracked quickly.

The problem with the secret key approach is that you must meet the person whom you are communicating with to exchange a key. You'd have to send a message containing information about the key, and that message could be compromised or eavesdropped upon. Quantum physics offers a solution.

Quantum keys

You could send your key using photons. A message in binary format – a string of 0s and 1s – can be passed along using photons with two polarizations, vertical and horizontal. And quantum uncertainty can be roped in to encrypt this information.

Imagine two people wishing to pass a message. Anne first carves her binary message onto a set of photons by setting their polarizations. To send the message privately she then scrambles the message. This can be done by sending the photons through a randomly chosen set of orthogonal filters, each able to measure two orthogonal polarization directions but oriented at 45 degrees to one another (+ or ×). Each

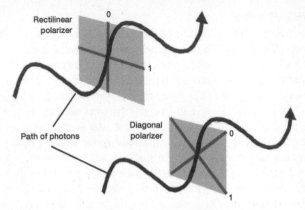

Rotated filters can be used to encrypt information in photons.

photon now has four possible polarization states – vertically, horizontally, angled left or angled right.

Anne's correspondent Bert now receives these scrambled photons. He too picks one filter for each and records what he measures. Thus far Bert just has a seemingly random set of observations. But the quantum magic comes when Anne and Bert compare notes. Bert tells Anne which filter he's used for each photon; Anne tells him if it is correct or incorrect. This information is enough for Bert to translate the binary message.

As only Bert knows the results, any third party can't figure out what the pair are talking about. Even better, if the eavesdropper tries to intercept the photons, quantum mechanics tells us that they will change the particles' properties. So Anne and Bert's comparison would throw up discrepancies – then they'd know that someone had been listening in.

Entangled messages

Quantum cryptography holds a lot of promise. But it is still a method firmly on the drawing board. Messages have been transmitted but over relatively short distances. The main problem is that any photon will interact with many other particles along the way and can lose its signal. A way around this information degradation is to employ quantum entanglement. An individual photon needn't run the gauntlet and travel kilometres to its destination – it is enough that the receiver has a coupled photon whose properties are entangled with the sender's partner particle. When the sender changes the state of her photon, the entangled partner simultaneously flips into a complementary state. So Bert could derive the message by adding a step that took into account quantum rules.

In 2007 Anton Zeilinger and his team in Austria managed to send messages over 144 km between two Canary Islands using entangled photon pairs – a feat known as quantum teleportation. The photons have opposite polarizations, set by coupling the particles at some point. Zeilinger's group managed to transmit information via an optical cable by manipulating one photon, and watching its entangled partner at the other end.

The condensed idea
Scrambled messages

46 Quantum dots

Tiny pieces of silicon and other semiconductors tens of atoms in size can act like a single molecule. Quantum effects come into play and all the electrons in the 'quantum dot' line up in energy according to quantum rules. Just as a hydrogen atom glows when its electrons jump down in energy, quantum dots may shine red, green and blue, making them useful as light sources and as biosensors.

From silicon chips to germanium diodes, much modern electronics is built around the semiconductor industry. Semiconducting materials normally don't conduct electricity – their electrons are locked within the crystal lattice. But if given an energy boost, electrons can be set free to roam within the crystal and carry a current.

The energy that the electrons need to gain to reach that mobility threshold is known as the 'band gap'. If the electrons exceed the gap energy then they become free to move, and the material's electrical resistance drops quite quickly. It is this flexibility – lying between insulators and conductors – that makes semiconductors so valuable for making controllable electronic devices. Most conventional electronic components use relatively large chunks of semiconducting material. You can place a silicon chip on your palm, or solder a new resistor into a radio. But in the 1980s physicists found that tiny pieces of these elements behave in unusual ways. Quantum effects kick in.

Tiny fragments of semiconducting elements like silicon, comprising just a few tens or hundreds of atoms, are known as 'quantum dots'. They are around a nanometre (a billionth of a metre) across, about the size of a large molecule. Because they are so small, the electrons within a quantum dot become correlated due to quantum connections. In essence, the whole ensemble starts to behave like one entity. They are sometimes known as 'artificial atoms'.

Because they are fermions, owing to the Pauli exclusion principle, each electron must occupy a different quantum state. A hierarchy of electrons results, giving the quantum dot a set of new energy levels, a bit like the many orbitals in a single atom. When an electron jumps up in energy, it leaves behind a 'hole' in the lattice, which relatively speaking is positively charged. The electron–hole pair is analogous to a hydrogen atom (a proton and an electron). And, like a hydrogen atom, the quantum dot can absorb and emit photons as electrons jump in energy. The quantum dot starts to glow.

The average energy spacing on the ladder of quantum states depends on the size of the dot. So the frequency of the light given off does likewise. Larger dots have more closely spaced energy gaps and glow red. Small dots glow blue. This opens up a range of uses for quantum dots as light sources, markers and sensors.

> The whole visible world is but an imperceptible speck in the ample bosom of Nature.
> Blaise Pascal, 1670

Dots at work

Physicists have long sought to make silicon emit light. Silicon is used in solar panels, for instance, because capturing ultraviolet light makes it start to conduct and causes electricity to flow. But the reverse seemed impossible, until in 1990 European researchers made a small piece of silicon glow red due to its quantum behaviour.

Since then, researchers have pushed silicon to glow green and blue. Blue light is especially valuable, as it is normally difficult to achieve without extreme laboratory conditions. Quantum dots may form the basis for new sorts of blue lasers.

Biosensors

Many biologists use chemical dyes, some fluorescent, to track changes in organisms during laboratory experiments or in the wild. Some have downsides. For example, they may deteriorate quickly and disappear or fade. Quantum dots offer some advantages. Because they are not chemically reactive, they persist for longer. And as the light they emit spans a narrow range of frequency it can be picked up more readily against the background through appropriate filters. Quantum dots can be tens of times brighter and a hundred times more stable than traditional dyes.

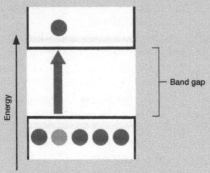
Silicon- and germanium-based dots now span the spectrum from infrared to ultraviolet wavelengths. Their luminosity can be tuned precisely and easily, simply by varying their size. Quantum dot technology can be used to make light-emitting diodes (LEDs), which are being pursued as a low-energy source of lighting as well as in television and computing screens. Dots may one day be used as qubits for quantum computing and cryptography. Because they act like individual atoms, they can even be entangled.

Quantum dots can be used as biosensors – detecting harmful chemicals or agents in the environment. They are longer-lived than fluorescent chemical dyes and emit light of a more exact frequency, making them easier to detect. Dots can also be used for optical technologies, such as very fast switches and logic gates for optical computing and for signalling down optical fibres.

How are quantum dots made? Most semiconductor devices are made by etching away a large sheet of material such as silicon.

Quantum dots are instead assembled atom by atom. Because they are built from the bottom up, their size and structure can be precisely controlled. Quantum dots may be grown as crystals in solutions. They can be produced in bulk, ending up as a powder or particles in a solution. As well as silicon and germanium, dots are often made from alloys of cadmium and indium. Some researchers are connecting up several quantum dots to make microscopic structures and circuits. The networks are linked by minuscule quantum wires. But the wires must be carefully made and attached to preserve the quantum state of the dot. They may be formed from long, thin organic molecules bonded chemically to the dot's surface. In this way lattices, sheets or other arrays of dots can be constructed.

> The history of semiconductor physics is not one of grand heroic theories, but one of painstaking intelligent labour.
> Ernest Braun, 1992

The condensed idea
All together now

47 Superconductivity

At supercold temperatures some metals, alloys and ceramics lose their electrical resistance completely. Currents are free to flow for billions of years without losing any energy. The reason is quantum mechanics. By pairing up, and with a gentle rocking from the positive ion lattice, electrons can stick together.

In 1911, the Dutch physicist Heike Kamerlingh Onnes was examining the properties of supercooled metals. He had worked out a way to cool helium to the point at which it becomes a liquid, a chilly 4.2 kelvins (above absolute zero, –273ºC, the lowest temperature possible). By bathing metals in liquid helium, he could investigate how their electrical behaviour altered. To his astonishment, when he placed a test tube of mercury into the liquid helium, the metal's electrical resistance plummeted. Mercury is normally a liquid at room temperature (around 300 kelvins); at 4 K it becomes solid. In this supercold state, mercury is perfectly conducting – its resistance is zero. Solid mercury is a 'superconductor'.

Other metals such as lead, niobium and rhodium were soon found to also be superconducting, although common materials used for wires at room temperature (copper, silver and gold) are not. Lead becomes superconducting at 7.2 K, and the other elements that do so each have a characteristic 'critical temperature' below which their resistance vanishes. The electrical currents that flow through superconductors never slow down. Currents can run through a supercooled lead ring for years without losing any energy. At room temperature, by contrast, they quickly decay. In superconductors the resistance is so low that currents could run for billions of years without weakening. Quantum rules prevent them from losing energy – there are no viable states by which they can do so.

Explaining superconductors

It took decades to find a full explanation for superconductivity. In 1957 US physicists John Bardeen, Leon Cooper and John Schrieffer published their 'BCS theory' of superconductivity describing how electron motions within a superconducting material become coordinated, so that they act as one system whose behaviour can be described using the equations of waves.

Metals are made up of a lattice of positively charged ions surrounded by a sea of electrons. The electrons are free to move around the lattice, producing electric current as they do so. But they must overcome forces that resist their movement. At room temperature, atoms are not still. They jiggle around. So moving electrons must dodge the jostling ions and may be scattered when they crash into them. These collisions produce electrical resistance, stalling the current and shedding energy. At supercold temperatures, the ions do not jiggle around as much. So the electrons can travel further without bouncing off. But this alone doesn't explain why the resistance drops suddenly to zero at the critical temperature, rather than declining slowly.

A clue as to what is going on is that the critical temperature scales with the atomic mass of the superconducting material. If it was simply due to electron properties that would not be the case, as all electrons are effectively the same. So heavier isotopes of mercury, for

> Thus the mercury at 4.2 K has entered a new state, which, owing to its particular electrical properties, can be called the state of superconductivity.
> Heike Kamerlingh Onnes, 1913

instance, have a slightly lower critical temperature. This hints that the entire metal lattice must be involved – the heavy ions are moving as well as the electrons.

The BCS theory supposes that the electrons hold hands and begin a sort of dance. The lattice's own vibrations give the electron waltz its timing. The electrons form loose pairs – known as Cooper pairs – whose motions are tied.

Electrons are fermions, which would normally be prevented from lying in the same quantum state by the Pauli exclusion principle. But when paired, the superconducting electrons behave more like bosons and can inhabit similar states. The ensemble's energy lowers as a result. An energy 'band gap' above them acts as a buffer. At very cold temperatures, the electrons don't have enough energy to break free and push through the lattice. So they avoid the collisions that cause resistance. The BCS theory predicts that superconductivity breaks down if the electrons acquire enough energy to jump the band gap. In agreement, the size of the band gap has been shown to scale with the critical temperature.

Magnetic levitation

If a small magnet is brought up to a superconductor, it will be repelled due to the Meissner effect. The superconductor essentially acts like a magnetic mirror, creating opposing fields on its surface that push the magnet away. This can lead to the magnet floating above the superconductor's surface – magnetic levitation. Such physics could be the basis for magnetic levitation, or 'maglev', systems for transport. Trains built on magnetic bases could hover and fly above superconducting rails without friction.

As well as having zero resistance, superconductors have another weird property – they cannot hold a magnetic field within them. This was discovered in 1933 by Walther Meissner and Robert Ochsenfeld, and is known as the Meissner effect. The superconductor expels magnetic fields by creating currents on its surface that exactly cancel what would be present inside it if it were a normal conductor.

Getting warmer

In the 1960s the race began to try to find new types of superconductors. Physicists wanted to find superconductors with high critical temperatures, which could be used more widely. Liquid helium is difficult to make and maintain. Liquid nitrogen, which sits at 77 K, is much easier to handle and produce. Physicists sought materials that could work at temperatures that can be reached with liquid nitrogen. Superconducting materials that work at room temperature are the ultimate goal, although we're still a long way from that.

Superconducting alloys, such as those of niobium and titanium and niobium and tin, were found to be superconducting at slightly higher temperatures (10 and 18 K respectively) than their raw elements. These were used to make superconducting wires, which were used to build strong magnets, which could be used for particle accelerators.

A further prediction by the British physicist Brian Josephson led to a string of new devices. Josephson worked out that current could be made to flow through a sandwich of two superconductors separated by a thin layer of insulator. The electric energy could quantum-tunnel through the sandwich filling – forming a Josephson junction. These are sensitive enough to measure tiny magnetic fields a billion times smaller than Earth's.

It takes a trained and discerning researcher to keep the goal in sight, and to detect evidence of the creeping progress toward it.

John C. Polanyi, 1986

In 1986, Georg Bednorz and Alex Müller discovered types of ceramics that could superconduct at 30 K, a big step forward. These were made from blends of barium, lanthanum, copper and oxygen (cuprates). This was unexpected because ceramics are usually used for insulators at normal temperature – as protectors on electrical pylons and substations, for instance.

A ceramic containing yttrium instead of lanthanum was found a year later that became superconducting at around 90 K. This broke the liquid nitrogen limit, making it economically feasible to use and ushering in a new scramble to find other high-critical-temperature superconductors. Today they have exceeded 130 K, but none is useful at room temperature.

The condensed idea
In the flow

48 Bose–Einstein condensates

When groups of bosons are extremely cold they can descend into their lowest energy state. There is no limit to how many bosons can hold the same state, and weird quantum-mechanical behaviours manifest themselves, such as superfluidity and interference.

Particles come in two types – bosons and fermions – according to whether they have integer or half-integer values of quantum spin. Bosons include photons, other force carriers and symmetric atoms such as helium (whose nucleus comprises two protons and two neutrons). Electrons, protons and neutrons are fermions.

According to Pauli's exclusion principle, no two fermions can exist in the same quantum state. Bosons, on the other hand, are free to do what they like. In 1924, Albert Einstein wondered what would happen if many bosons got together into a single ground state, as if they were crushed into a quantum black hole. How would this community of clones behave?

Satyendra Nath Bose, an Indian physicist, had sent Einstein a paper on the quantum statistics of photons. Einstein thought the work so important that he translated and republished Bose's paper in German, and set to work extending the ideas to other particles. The result was a statistical description of the quantum properties of bosons, which are now named after Bose. Bose and Einstein imagined a gas made of bosons. Just as atoms in a vapour take on a range of energies around an average speed that depends on the temperature of the gas, so the bosons would adopt a range of quantum states. The physicists derived a mathematical expression for this distribution of states, now known as Bose–Einstein statistics, which applies to any group of bosons.

Einstein then asked what would happen if the temperature was dropped. The bosons would all sink in energy. Eventually, he reasoned, most of them could 'condense' into the lowest possible energy state. In theory, an indefinite number could sit at this minimum energy, forming a new type of matter that we now call a Bose–Einstein condensate. If made of many atoms, condensates could exhibit quantum behaviour on a grand scale.

Superfluids

Creating a Bose–Einstein condensate gas in the laboratory had to wait until the 1990s. Meanwhile, insights came from studies of helium. Liquid helium condenses at a temperature of around 4 kelvins. As Pyotr Kapitsa, John Allen and Don Misener discovered in 1938, if helium is cooled even further to around 2 kelvins, it starts to behave very strangely. Just as supercooled mercury suddenly becomes superconducting, liquid helium starts to lose its resistance to flow.

The liquid helium becomes a 'superfluid' with zero viscosity. Fritz London proposed Bose–Einstein condensation as a possible mechanism for this strange behaviour – some of the helium atoms have dropped down collectively into the lowest energy state, where they are not susceptible to collisions. But, being a liquid rather than a gas, superfluid helium doesn't quite fit Einstein's equations well enough to test London's proposal.

It took a long time for physicists to develop the technologies needed to make a gaseous condensate in the laboratory. Getting so many particles into one quantum state isn't easy. The particles involved must be quantum-mechanically identical, which is hard to achieve for entire atoms. The best way forward is to make a dilute gas of atoms, cool them to extremely low temperatures, and bring them close enough together so that their wavefunctions overlap.

Atoms can now be cooled, by holding them within magnetic traps and firing lasers at them, to temperatures of just billionths of a kelvin (nanokelvins). In 1995 Eric Cornell and Carl Wieman at the University of Colorado at Boulder managed to create the first Bose–Einstein condensate using around 2,000 rubidium atoms at just 170 nanokelvins.

> From a certain temperature on, the molecules 'condense' without attractive forces, that is, they accumulate at zero velocity. The theory is pretty, but is there also some truth to it?
>
> Albert Einstein, 1924

A few months later, Wolfgang Ketterle at MIT, who later shared the Nobel Prize with Cornell and Wieman, followed suit with sodium atoms. By using a hundred times more atoms, Ketterle was able to reveal new behaviour such as quantum interference between two condensates.

Supercold strangeness

There is now a great deal of research being done on Bose–Einstein condensates and superfluids, which is revealing their strange properties. When condensates and superfluids are stirred or set in rotation, vortices or whirlpools can emerge. The angular momentum of these eddies is quantized, coming in multiples of a basic unit.

When condensates grow too large they become unstable and explode. Bose–Einstein condensates are thus very fragile. The slightest interaction with the outside world, or any warming, can destroy them. Experimenters are exploring ways to stabilize atoms so that larger condensates can be assembled.

One factor is the natural attraction or repulsion of atoms. Lithium atoms, for instance, tend to attract one another. So condensates built from them suddenly implode when they reach some threshold size, simultaneously blasting out most of the material, just like a supernova explosion. Isotopes of atoms that naturally repel one another, such as rubidium-87, can be used to build more stable condensates.

Condensates and superfluids can be used to slow light to a halt. In 1999 the Harvard University physicist Lene Hau brought a beam of laser light to a crawl, and later to a complete stop, by shining it through a glass cell filled with ultracold sodium vapour. The condensate effectively tries to pull the incoming photons into its state, dragging on them until they stop.

Hau turned down the laser's brightness, until there were no photons left in the condensate. Yet the spins of the photons remained imprinted in the sodium atoms. This quantum information can then be freed by shining another laser beam through the cell. Information can thus be transmitted by light but stored on and retrieved from ultracold atoms. So Bose–Einstein condensates might be used for quantum communication one day.

The condensed idea
Human wrongs?

49 Quantum biology

Quantum effects such as wave–particle duality, tunnelling and entanglement may play important roles in living organisms. They make chemical reactions work, channel energy around cells and may tell birds how to navigate using Earth's magnetism.

Quantum mechanics governs the cold and probabilistic world of the atom. But how important is it in the natural world? On the one hand, quantum mechanics must operate to some extent at the level of individual molecules in plants or the animal or human body. But it's difficult to imagine how quantum wavefunctions remain coherent among the messy goings-on of a cell or in a bacterium.

> Chromosome structures . . . are law-code and executive power – or to use another simile, they are architect's plan and builder's craft – in one.
>
> Erwin Schrödinger, 1944

The Austrian physicist Erwin Schrödinger was one of the first to discuss quantum biology in his 1944 book *What Is Life?* Scientists today are making discoveries that suggest that quantum mechanics does play a part in important natural phenomena. Birds may apply their quantum skill to sense Earth's magnetic field and use it for navigation. Photosynthesis, the vital process by which organisms convert sunlight, water and carbon dioxide into fuel, may also hinge on subatomic processes. When sunlight hits a leaf, photons crash into chlorophyll molecules. The chlorophyll absorbs the photon's energy, but must then channel that energy towards the chemical factory part of the cell that is busy making sugars. How does the cell know how to do this efficiently?

The photon's energy spreads as waves across the plant cell. Just as the theory of quantum electrodynamics describes interactions between photons and matter in terms of combinations of all possible paths, the most probable route being the outcome, so the transmission of energy through the leaf cell can be described as a superposition of waves. In the end, the optimal pathway draws the photon's energy to the chemical reaction centre of the cell.

Chemists at the University of California, Berkeley, and elsewhere have found experimental evidence to back up this idea in recent years.

By firing laser pulses at photosynthesizing cells within bacteria, they identified waves of energy flowing across the cell. These waves behave in concert, and even exhibit interference effects, proving that they are coherent. All this takes place at normal ambient temperatures.

It is an open question as to why these coordinated quantum effects are not quickly disrupted by all the chemical goings-on within cells. The chemist Seth Lloyd has suggested that random noise within the cell environment can actually help the photosynthesis process along. All the jostling stops the wave energy from becoming locked at particular sites, gently rocking it free.

Quantum sensing

Quantum effects are also important in other reactions within cells. Quantum tunnelling by protons from one molecule to another is a feature of some reactions catalysed by enzymes. Without the addition of a helping hand from quantum-mechanical probability, the proton should not be able to jump the energy barrier required. Electron tunnelling might also lie behind our sense of smell, explaining how receptors in our nose pick up biochemical vibrations.

What is Life?

In 1944 Erwin Schrödinger published a popular science book called *What Is Life?* In it he summarized the lessons physics and chemistry held for biology, based on a series of public lectures he gave in Dublin. Schrödinger believed that hereditary information was held in a molecule, stored in its chemical bonds. (Genes and the role of DNA in reproduction were unknown at the time.) The book begins by explaining how order comes out of disorder. Because life requires order, the master code of a living organism needs to be lengthy, made of lots of atoms, able to be arranged. Mutations come from quantum leaps. The book concludes with his musings on consciousness and free will. Schrödinger believed that consciousness is a state that is separate from the body, although dependent on it.

Migrating birds take their cues from Earth's magnetic field. Photons striking a bird's retina activate a magnetic sensor. The mechanism by which this happens isn't precisely known, but one possibility is that the incoming photon creates a pair of free radicals – molecules with a single electron outstanding, which makes it react easily with other molecules. The quantum spin of these extra lone electrons may align with the magnetic field.

> From all we have learnt about the structure of living matter, we must be prepared to find it working in a manner that cannot be reduced to the ordinary laws of physics.
> Erwin Schrödinger, 1956

The molecules react with others in different ways depending on the electron's spin, so conveying the direction of the geomagnetic field. Some chemical is made if the system is in one state, but not when it is in the other. So the concentration of the chemical can communicate to the bird the direction of Earth's magnetism.

Simon Benjamin, a physicist at Oxford University, has proposed that the two single electrons attached to the free radicals could also be entangled. So if the molecules become separated, their quantum spin states remain linked. Researchers have suggested that entanglement could be maintained for tens of microseconds in a bird's inner compass, much longer than in many 'warm and wet' chemical systems.

Quantum mechanics could assist other animals and plants with directional sensing. Some insects and plants are sensitive to magnetic fields. For example, the growth of the flowering plant *Arabidopsis thaliana* is inhibited by blue light, but magnetic fields can modify that effect, perhaps also involving the radical-pair mechanism.

Quantum skill confers many advantages to organisms. It seems to overcome nature's tendency for disorder and operates at ambient temperatures, unlike many situations in physics that require extreme supercooled environments.

The question of how or whether such skills evolved is unanswered. Scientists don't know if quantum effects are favoured by natural selection, or if they are an accidental by-product of the close-packed systems from which organisms are formed. One day it might be possible to compare molecules from species of algae, for instance, that evolved at different times to look for evolutionary changes over time.

Should scientists learn more about quantum effects in organisms, they could generate exciting new technologies. Artificial photosynthesis could be a radical new power source, perhaps leading to very efficient new forms of solar cells. Quantum computing might also benefit from understanding how biological systems avoid decoherence.

The condensed idea
A little helping hand

50 Quantum consciousness

From free will to our sense of time, there are parallels between how our minds work and quantum theory. Many physicists have wondered if that means there is a deep link. Speculation is rife about whether we might experience consciousness due to quantum tickling of microscopic structures in our brain, collapsing wavefunctions or entanglement.

With its tangled networks of neurons and synapses, the brain is one of the most complex systems known. No computer can match its processing power. Could quantum theory explain some of the brain's unique qualities?

There are two main differences between the brain and a computer – memory and processing speed. A computer has a far bigger memory than the brain – a hard disk can be almost infinitely large. But the brain wins hands down at speed learning. Humans can spot a person in a crowd a lot quicker than any automaton can.

The brain's processing power is hundreds of thousands of times greater than the most advanced computer chips. Yet signals in the brain are transmitted at a relative snail's pace – up to six orders of magnitude slower than digital signals. As a result of these different speeds, the brain has a hierarchical structure, built up from many layers that talk to one another. Computers have essentially one layer, churning through millions of calculations to beat human grandmasters at chess, for instance.

Consciousness

How might computations in the brain give rise to consciousness? It's hard to define what consciousness is exactly. But it is how we experience life. We have a sense of the present – living in the now. And we have a sense of the passage of time – the past. Our brain stores memories, and we assign patterns to them to give meaning. We can make simple predictions about the future, through which we make decisions.

Many physicists, including the quantum pioneers Niels Bohr and Erwin Schrödinger, have thought that biological systems, including brains, might behave in ways that are indescribable using classical physics. As quantum theory developed, a number of ways of creating consciousness have been proposed, from collapsing wavefunctions to entanglement. But we are still far from learning exactly how this works.

David Bohm asked what happens when we listen to music. As the tune rolls along, we retain a memory of its evolving shape and combine that recollection with our sensory experience of the present, the sounds, chords and feelings of the music we are hearing now. It is this blend of the historical pattern with our canvas of the present that is our experience of consciousness.

Bohm argued that this coherent narrative stems from the underlying order of the universe. Just as photons are both waves and particles and we observe one form under different circumstances, so mind and matter are projections onto our world of a deeper order. They are separate aspects of life: being complementary, looking at matter tells us nothing about consciousness, and vice versa.

Artificial intelligence

One of the first people to try to quantify how the brain handles information was the British mathematician Alan Turing. Now recognized as the father of computing, in 1936 he published a famous paper proving that it was possible to build a machine to handle any calculation that could be expressed as a series of rules, an algorithm. He tried to imagine the brain as a sort of computer, and wondered by what rules it worked. Turing proposed a test of artificial intelligence, now known as the Turing test, as follows. A computer could only be considered intelligent if it could answer any question in such a way that it could not be distinguished from a human.

In 2011, a computer called Watson came close. On the US television quiz show *Jeopardy!* the machine beat two human contestants, making sense of a host of English-language colloquialisms, metaphors, puns and jokes to bag the prize. Watson was proof of a concept for artificial intelligence researchers. But its logical system is quite unlike the human brain.

Quantum brain states

In 1989, the Oxford mathematician and cosmologist Roger Penrose published one of the most controversial ideas for how consciousness arises in *The Emperor's New Mind*. Penrose revisited Turing's ideas and argued that the human brain is not a computer. Moreover, the way it operates is fundamentally different and no computer could ever replicate it using logic alone.

Penrose went several big steps further, by proposing that consciousness is linked to fluctuations in space-time due to quantum gravity. Most physicists didn't like it – why should quantum gravity apply to a soft, wet, gelatinous brain? The artificial intelligence community didn't like it, as they believed they would one day build a powerful brain simulator.

> I regard consciousness as fundamental. I regard matter as derivative from consciousness. We cannot get behind consciousness. Everything that we talk about, everything that we regard as existing, postulates consciousness.
>
> Max Planck, 1931

Penrose didn't know exactly how or where the brain handled these quantum gravity effects. He teamed up with the anaesthesiologist Stuart Hameroff to extend the model, elaborated in Penrose's 1994 book *Shadows of the Mind*. The conscious mind, they suggested, was made up of many superposed quantum states, each with its own space-time geometry. The states decay as events unfold, but they don't all do so instantaneously. This momentary awareness is our feeling of consciousness.

Quantum gravity acts at very tiny scales, smaller than a neuron. Hameroff suggested that this could take place in long tubular polymer structures that lie within neurons and other cells, called microtubules. Microtubules provide scaffolding, and they also shepherd neurotransmitting chemicals.

Bose–Einstein condensates, wavefunction collapse and the interface between the observer and the observed have been explored as consciousness triggers. And quantum field theory has also been explored as a means of describing brain states. Memory states may be described as many-particle systems, a bit like the virtual sea of particles that are associated with quantum fields and empty space. Quantum tunnelling may help along the chemical reactions involved in neuronal signalling.

Others physicists have suggested that quantum randomness underlies consciousness, jolting us sequentially from one mindful state to another. Many physicists remain sceptical, though, and have queried whether quantum states could exist in the brain for any length of time. In a 1999 paper, the physicist Max Tegmark suggested that decoherence effects would dissemble quantum states on a timescale much shorter than that characteristic of brain signalling. The brain is too big and hot to be a quantum device. So the jury is definitely out on the degree to which quantum theory explains consciousness.

The condensed idea
Mental collapse

Glossary

Antimatter: a complementary state to normal matter with quantum parameters reversed.

Atom: the smallest piece of matter that can exist independently; atoms comprise a nucleus (of protons and neutrons) surrounded by electrons.

Baryon: elementary particle (such as a proton) made of three quarks.

Black-body radiation: characteristic glow of a perfectly dark substance.

Boson: particle with integer spin, such as a photon.

Complementarity: argument that the nature of a quantum phenomenon depends on the way in which it is measured.

Cosmic microwave background: weak microwave glow coming from across the sky, originating in the early universe.

Cosmology: the study of the history of the universe.

Electromagnetism: theory that unifies electricity and magnetism.

Electron shells: regions of space where electrons can be found surrounding the atomic nucleus.

Energy: the potential something has for change; it is conserved overall.

Entanglement: correlated signals between particles.

Fermion: particle with a half-integer spin; no two fermions can have the same quantum state.

Field: how force is transmitted over distance.

Fission: the splitting of a large nucleus.

Force: a push, pull or other impulse that causes something to change position.

Frequency: the rate at which wave peaks pass a point.

Fusion: the joining together of small nuclei.

Gravity: a force by which masses attract.

Hadron: elementary particle made of quarks (baryons and mesons are subclasses).

Interference: the reinforcement or cancellation of waves when combined.

Isotope: versions of a chemical element with differing numbers of neutrons.

Locality: principle that an object is influenced only by its immediate surroundings.

Many worlds hypothesis: the idea that quantum events cause parallel universes to branch off.

Mass: a property of bulk that depends on how many atoms or how much energy an object contains.

Matrix: a mathematical construct similar to a table of numbers.

Momentum: product of mass and velocity.

Molecule: two or more atoms joined together by bonds.

Nucleus: the compact core of the atom, comprising protons and neutrons.

Observer: in quantum mechanics, the witness to a measurement.

Phase: the relative difference between two waves, measured as a fraction of wavelength.

Photon: a particle of light.

Quanta: packets of energy.

Quark: the smallest constituent of a hadron, such as a proton or neutron.

Qubits: 'quantum bits', elements of quantum information.

Radioactivity: the emission of particles by unstable nuclei.

Randomness: an outcome that is purely based on chance, such as throwing dice.

Semiconductor: a material that conducts electricity more than an insulator but less than a conductor.

Space-time: combination of three dimensions of space and one of time in relativity theory.

Spectrum: the brightness of light at a range of frequencies.

Superconductivity: conduction of electricity without any resistance.

Superfluid: motion of a liquid with no viscosity.

Symmetry: similarity under reflection or rotation or re-scaling.

Uncertainty: in quantum mechanics, the idea that some quantities cannot be known simultaneously.

Universe: all space and time; physicists' descriptions may go beyond this when talking about parallel universes and string theory.

Vacuum: empty space.

Wave function: in quantum theory, a wave-like probability function that describes a particle's properties.

Wavelength: The distance between wave crests or troughs.

Wave–particle duality: idea that quantum entities such as light can appear as either particles or waves (see 'complementarity').

Index

About the author

Joanne Baker studied Natural Sciences at Cambridge and took her PhD at the University of Sydney. She is currently Senior Comment Editor at *Nature* magazine.

Greenfinch,
An imprint of Quercus Editions Ltd
Carmelite House
50 Victoria Embankment
London
EC4Y 0DZ

First published in 2013

A catalogue record of this book is available from the British Library

Designed and illustrated by Patrick Nugent

ISBN 978 1 52942 930 5
eBook ISBN 978 1 52942 929 9

Printed and bound in Great Britain by Clays Ltd, Elcograf S.p.A.

10 9 8 7 6 5 4 3 2 1

MIX
Paper from
responsible sources
FSC® C104740

Papers used by Greenfinch are from well-managed forests and other responsible sources.